Nanotechnology to Aid Chemical and Biological Defense

NATO Science for Peace and Security Series

This Series presents the results of scientific meetings supported under the NATO Programme: Science for Peace and Security (SPS).

The NATO SPS Programme supports meetings in the following Key Priority areas: (1) Defence Against Terrorism; (2) Countering other Threats to Security and (3) NATO, Partner and Mediterranean Dialogue Country Priorities. The types of meeting supported are generally "Advanced Study Institutes" and "Advanced Research Workshops". The NATO SPS Series collects together the results of these meetings. The meetings are co-organized by scientists from NATO countries and scientists from NATO's "Partner" or "Mediterranean Dialogue" countries. The observations and recommendations made at the meetings, as well as the contents of the volumes in the Series, reflect those of participants and contributors only; they should not necessarily be regarded as reflecting NATO views or policy.

Advanced Study Institutes (ASI) are high-level tutorial courses to convey the latest developments in a subject to an advanced-level audience

Advanced Research Workshops (ARW) are expert meetings where an intense but informal exchange of views at the frontiers of a subject aims at identifying directions for future action

Following a transformation of the programme in 2006 the Series has been re-named and re-organised. Recent volumes on topics not related to security, which result from meetings supported under the programme earlier, may be found in the NATO Science Series.

The Series is published by IOS Press, Amsterdam, and Springer, Dordrecht, in conjunction with the NATO Emerging Security Challenges Division.

Sub-Series

A.	Chemistry and Biology	Springer
B.	Physics and Biophysics	Springer
C.	Environmental Security	Springer
D.	Information and Communication Security	IOS Press
E.	Human and Societal Dynamics	IOS Press

http://www.nato.int/science
http://www.springer.com
http://www.iospress.nl

Series A: Chemistry and Biology

Nanotechnology to Aid Chemical and Biological Defense

edited by

Terri A. Camesano
Worcester Polytechnic Institute, Worcester, MA, USA

Springer

Published in Cooperation with NATO Emerging Security Challenges Division

Proceedings of the NATO Advanced Research Workshop on Nanotechnology to Aid Chemical and Biological Defense
Antalya, Turkey
22–26 September 2014

Library of Congress Control Number: 2015949813

ISBN 978-94-017-7244-0 (PB)
ISBN 978-94-017-7217-4 (HB)
ISBN 978-94-017-7218-1 (e-book)
DOI 10.1007/978-94-017-7218-1

Published by Springer,
P.O. Box 17, 3300 AA Dordrecht, The Netherlands.

www.springer.com

Printed on acid-free paper

For Nagu

Preface

The NATO Science for Peace and Security (SPS) Programme has identified the need to "facilitate mutually beneficial cooperation on issues of common interest, including international efforts to meet emerging security challenges". An Advanced Research Workshop (ARW) was held on "Nanotechnology to Aid Chemical and Biological Defense" on September 22–26, 2014, in Antalya, Turkey, to address the NATO SPS key priority of *Defense against CBRN Agents and Environmental Security*. The ARW was co-organized by researchers from the USA and Georgia. Speakers and other participants came from the USA, Georgia, Italy, Germany, Belgium, France, Turkey, Croatia, UK, Slovenia, Ukraine, Moldova, Romania and Kyrgyz Republic. This book is representative of the scientific contributions made at this meeting emphasizing both fundamental research and practical implications of the research to CBRN defense and environmental security.

In Chap. 1 the potential of Atomic Force Microscopy (AFM) to characterize the nanoscale properties of pathogens such as *Candida*, *Aspergillus* and *Staphylococcus* species is elaborated, with AFM operated in various modes. Microbial pathogens are highly complex and heterogeneous systems, exhibiting differences in growth rate, as well as resistance to stress and drug treatment. The authors show that using topographic imaging, one can visualize the ultrastructure of live cells and their subtle modification under activity of antimicrobial agents. AFM can be operated under force spectroscopy mode with tips functionalized by biomolecular ligands to probe the localization and adhesion of single receptors on cells, such as cell adhesion proteins and antibiotic binding sites. They show that single-cell force spectroscopy can be used to quantify the forces driving microbe-microbe, microbe-solid and microbe-host interactions.

Chapter 2 summarizes experiences in the development of bacteriophage-based therapeutics, prophylactic and diagnostic preparations and their uses in different fields, such as medicine, veterinary, agriculture, food and water safety, etc. Bacteriophages are obligate intracellular parasites that are highly specific to their host pathogens. With the rise of antibiotic-resistant organisms, bacteriophages are beginning to be explored to target and treat multidrug-resistant pathogenic bacteria. The nanoscale of bacteriophages and the ease of implementing genetic modifications have made bacteriophages attractive biosensors for pathogen detection with high levels of reliability, sensitivity and selectivity with short assay times. Bacteriophage-based therapeutics which is still at early stages of development is discussed in this chapter. An important application of bacteriophages to environmental and thereby economic security is addressed in Chap. 3. It is shown that bacteriophages specific to *Xanthomonas vesicatoria* strains spread in Georgia can be used to efficiently prevent tomato bacterial spot in laboratory conditions under artificial infection. This finding and approach are of critical significance to the agricultural sector in many countries. Chapter 4 describes phage display as a way for identifying monoclonal antibodies for diagnostic purposes, especially in dealing with toxic and non-immunogenic antigens. The author shows that panning large phage libraries can be performed directly on whole cells to identify antibodies that will recognize their membrane-bound antigens in their native conformation and lipid environment. This opportunity is particularly meaningful when it is necessary to isolate antibodies that are able to bind accessible epitopes in vivo, as it is the case of biomarkers exposed at the cell membrane surface of pathogenic microorganisms. The author also suggests strategies useful to overcome common drawbacks of phage display technology.

The authors of Chap. 5 are concerned with environmental security threat posed by contamination of water sources by nanomaterials. They have conducted studies using model membranes to explore whether nanoparticles disrupt the membrane structure which would make them cytotoxic and what characteristics of nanoparticles may determine the mode and mechanisms of their interactions with membrane systems. Of special interest is the use of quartz crystal microbalance with dissipation monitoring as a technique to extract nanoscale information regarding nanoparticle-membrane interactions. Chapter 6 advances in detail an inexpensive bioassay suited for assessing chemical poisoning in the environment such as the presence of pesticides and other organophosphates, metal contaminants or nanoparticles. The authors show that their test with terrestrial isopods, a single species test, is flexible in terms of exposure duration and biomarker selection. By alternating test duration, one can assess the toxic potential of substances as low, moderate or high, while a variety of biomarkers at different levels of biological complexity increase the relevance of test results. Another instrumental approach, the electroanalytical method, to look at the toxicity of nanomaterials in the marine environment is described in Chap. 7. The method is shown to be advantageous for studying different biogeochemical processes in the marine environment, especially those related with organic matter, sulphur species and trace metals cycling and the interaction and distribution between dissolved and colloidal phases.

Chapter 8 describes a novel procedure based on metal oxide nanotubes to detect ultra-trace quantities of the explosive pentaerythritol tetranitrate (PETN). The inexpensive, lightweight, easily made and cost-effective devices are shown to detect PETN down to about 112 ppt. The authors provide a step-by-step description for the synthesis of anatase titanium dioxide nanotubes grown by electroless deposition under aqueous conditions in ion track-etched polycarbonate templates. Details of sensor construction on a chip are also provided. The synthesis of another type of nanomaterial, graphite-like carbon nitride, that can be employed as a catalyst for agent inactivation or as a chemical sensor is described in Chap. 9. Carbon nitride is a unique material because of the possibility to dope its structure by non-metal atoms though which one can impart novel chemical and electronic characteristics. The authors describe their development of a new method of pyridine pyrolysis to obtain O-doped graphite-like carbon nitride with high contents of oxygen. The authors of Chap. 10 report on citrate-capped, colloidal gold nanoparticle (AuNP) film assemblies of varying particle sizes (5–50 nm) adsorbed to bulk Au substrates to serve as platform electrochemical sensors to simultaneously detect and reduce the explosive trinitrotoluene (TNT) to 2,4,6-triaminotoluene (TAT) in solution. The high surface area-to-volume ratio of colloidal AuNPs offers advantages in electrocatalysis and enhanced signal transduction, while the facile immobilization onto a variety of substrates provides an adaptable and reproducible platform technology.

Material advances in the synthesis of chalcogenide material are discussed in Chap. 11. These novel class of materials are good candidates for photonic and optoelectronic applications such as optical elements and memories, optical sensors, nonlinear optical devices, holographic elements, IR telecommunications, biosensing and signal processing applications because they possess high strength/hardness, excellent wear/corrosion resistance and excellent optical properties (high refractive index, high transmittance in the near IR and IR regions of the spectrum). The report focuses on the characterization of chalcogenide incorporating tin and describes changes in measured properties as a function of tin content. In Chap. 12 another important type of porous material, zeolitic imidazolate frameworks (ZIFs), is described. They are promising materials for use in the capture and/or detection of hazardous chemicals under humid environments. In this work molecular simulations have been carried out to find a suitable ZIF structure to handle one explosive (nitromethane), six toxic chemicals (hydrogen disulphide, sulphur dioxide, nitrogen dioxide, carbon monoxide, ethylene oxide, benzene) and three warfare agents (sarin, sulphur mustard, phosgene oxime). Super-resolution imaging using fluorescent labels to image nanoscale biomolecular structures is discussed in Chap. 13. The authors propose the development of a super-resolved stochastic hyperspectral Raman microscopy technique for imaging of biological architectures. The surface-enhanced Raman spectroscopy (SERS) signal contains information about the presence of various Raman bands, allowing for the discrimination of families of biomolecules such as lipids, proteins and the DNA. The last chapter chronicles the lingering effect of radioactive wastes in soil and water in the former Soviet Republic, Kyrgyzstan. The paper presents an assessment of the radioactive material threat in soil and water sources in all parts of the country.

The Advanced Research Workshop also included a few presentations that are not included in this book: cantilever-based sensors for detecting *E. coli* O157:H7, waterborne parasites, food and water toxins and *B. anthracis*; nanomaterials-based chemical sensors, many utilizing metal oxides in the form of thin films, nanoparticles and nanotubes; and the development of gas sensors based on nanowire bundles exploiting both the chemiresistor and surface ionization configurations.

The Advanced Research Workshop brought together a diverse group of international civilian researchers focused on nanoscience and nanotechnology problems that are relevant to chemical and biological defense needs, in order to share current state of the art and discuss future opportunities and challenges. The workshop and this resulting book have provided many examples of nanomaterials and nanotechnology having the potential to be used for detection and decontamination of chemical and biological threat agents and in the development of protective technologies. These capabilities are of increasing importance to international security, as chemical and biological agents begin to be used more frequently in acts of terrorism, in ways in which we have not previously been able to respond effectively. Cooperative research among NATO countries and NATO partners, as represented in this Advanced Research Workshop and this book, can make a critical contribution to meeting these challenges.

Worcester, MA, USA Terri A. Camesano

Acknowledgements

The chapters in this work are based on presentations made at an Advanced Research Workshop on "Nanotechnology to Aid Chemical and Biological Defense", which was held September 22–26, 2014, in Antalya, Turkey. Support for the ARW was largely provided by the NATO Science for Peace and Security Programme. Some additional funds were also provided by the Army Research Office.

The editor would like to thank all of the authors for contributing their fully engaged participation in the workshop and for contributing these excellent chapters. In addition, she thanks Ms. Felicia Vidito and Ms. Tiffany Royal from Worcester Polytechnic Institute for their dedicated assistance in organizing the logistics of the workshop.

Contents

Chapter 1
Atomic Force Microscopy Tools to Characterize the Physicochemical and Mechanical Properties of Pathogens

Sofiane El-Kirat-Chatel and Audrey Beaussart

Abstract Microbial pathogens are highly complex and heterogeneous systems. Cell populations generally contain subgroups of cells which exhibit differences in growth rate, as well as resistance to stress and drug treatment. In addition, individual cells are spatially organized and heterogeneous; this cellular heterogeneity is used to perform key functions. This complexity emphasizes the need for single-cell analysis techniques in microbial research.

With its ability to image and manipulate cellular systems at nanometer resolution and in physiological conditions, atomic force microscopy (AFM) offers unprecedented opportunities in microbiology and contributes to the birth of a new field called 'microbial nanoscopy'. Using topographic imaging, researchers can visualize the ultrastructure of live cells and their subtle modification under activity of antimicrobial agents. Force spectroscopy with tips that bear bioligands offers a means to probe the localization and adhesion of single receptors on cells, such as cell adhesion proteins and antibiotic binding sites. Single-cell force spectroscopy quantifies the forces driving microbe-microbe, microbe-solid, and microbe-host interactions. In this chapter, we will discuss how we can use these AFM modalities in microbiology. We will present some recent breakthroughs in pathogen research, emphasizing the potential of various AFM modes for studying cell adhesion and biofilm formation in *Candida, Aspergillus and Staphylococcus* species.

1.1 Introduction

Candida, *Aspergillus* and *Staphylococcus* are respectively fungal and bacterial genus and contain species frequently isolated as human pathogens. These microorganisms can be commensal and are mainly opportunistic pathogens, i.e., they are

S. El-Kirat-Chatel (✉) • A. Beaussart (✉)
Université Catholique de Louvain, Institute of Life Sciences, Croix du Sud, 4-5, bte L7.07.06, 1348 Louvain-la-Neuve, Belgium
e-mail: sofiane.elkirat@uclouvain.be; audrey.beaussart@uclouvain.be

T.A. Camesano (ed.), *Nanotechnology to Aid Chemical and Biological Defense*,
NATO Science for Peace and Security Series A: Chemistry and Biology,
DOI 10.1007/978-94-017-7218-1_1

non-pathogenic for healthy individuals or responsible of minor and superficial infections, however, when ones immune system is depressed, they could invade patients leading to severe diseases. One of the most important factors of the pathogenicity of these microbes is their capability to adhere and subsequently form biofilm on abiotic and biotic surfaces [1–7]. Elucidating the molecular mechanisms that govern their adhesion is then crucial for the development of new antimicrobial strategies.

Candida species are at the origin of numerous mucosal infections which represent an important clinical and economical problem [8, 9]. Among *Candida*, various species are emerging as important pathogens, including *C. glabrata*, C. *tropicalis*, *C. krusei*, *C. parapsilosis* and *C. lusitaniae* [9, 10]. The most frequently isolated *Candida* species, *Candida albicans*, is threatened by very few antifungal treatments, as the arsenal available today consists of only four drug classes [11–13]. One of the numerous virulence factors of *C. albicans* is the polymorphic switch from budding yeast to a filamentous hyphae (so-called morphogenesis) which facilitates adhesion and biofilm formation [2, 3, 14]. This pathogenic yeast exposes at its surface compounds that are grouped as pathogen associated molecular patterns (PAMPs) that will be recognized by receptors of cells from our innate immunity [14, 15], more specifically of macrophages and neutrophils phagocytes [16]. These specific receptors are named pattern recognition receptors (PRRs) [16]. Additionally, *C. albicans* cell wall contains adhesins grouped as Als (Agglutinin like sequence) that promote adhesion to various abiotic and biotic surfaces.

Also of importance for public health is *Aspergillus fumigatus*, a ubiquitous saprophytic mold species which can cause a variety of pathologies in humans [17]. The cell wall of *A. fumigatus* acts as a sieve and reservoir for the molecules, which are the key players during infections [18]. As for *C. albicans*, *A. fumigatus* also exists as different morphotypes whose cell wall compositions largely differ. Airborne conidia are covered by a layer of hydrophobic organized proteins, called rodlets, partially imparting immunological inertness to the fungi [19]. Upon germination, the hyphae will expose different surface compounds such as galactosaminogalactan, a polysaccharide recently identified as a major virulence factor [20].

Finally, we will focus on the *Staphylococcus* species. *S. epidermidis* and its more virulent cousin *S. aureus* rank among the most causative agents of nosocomial infections [6]. *S. aureus* can cause potentially life-threatening infections such as sepsis, endocarditis and pneumonia [4], whereas *S. epidermidis* represents the most common source of infections on indwelling medical devices such as catheters and prostheses [6]. Host attachment of these bacteria is driven by a family of adhesive proteins, the Microbial Surface Components Recognizing Adhesive Matrix Molecules (MSCRAMMs), which are directly in contact with the host tissues and subjected to pressure of host defenses [4, 6, 21]. Additionally, antibiotic treatments are becoming more and more obsolete with the emergence of new antibiotic-resistance strains [4, 22].

Getting a complete understanding of the cell wall composition of the pathogens and their adhesion mechanisms is then crucial to decrease their biofilm formation and help the immune system to eradicate them. In this chapter, we will explain how

AFM tools can help decipher the adhesive and pathogenic properties of the species *C. albicans*, *A. fumigatus*, *S. aureus*, *S. epidermidis* and analyze their behavior when subjected to potential drugs. Technical advances presented here could be applied to a variety of other microbial cells.

1.2 General Principle of AFM

AFM has recently evolved as a multiparametric analysis tool which enables the imaging of the surface of living cells, as well as the detection and manipulation of surface single molecules. The application of this versatile instrument to cell biology has revolutionized our understanding of breakthrough mechanisms at the cell surface, providing a novel view of the structure-function relationship of surface components.

AFM has the powerful ability to operate in liquid medium, enabling observation of microorganisms with a nanometre resolution in physiological conditions. The principle of AFM is simple: a sharp tip mounted on a soft cantilever scans a defined surface area while the force interacting with the tip is recorded. The force is quantified via the deflection of a laser reflected on the cantilever apex and reflected in a photodiode. The three dimensional (x, y, z) precise positioning of the sample is ensured by a piezoelectric scanner. The extreme sensitivity of the instrument allows to measure forces in the piconewton range.

AFM imaging has first allowed biologists to observe structural composition of the microbes surface and to follow their evolution under the effect of external stimuli, such as temperature or drug injection. In addition, AFM can be used as a sensor probing the forces between the tip and the sample. If a single molecule is attached to the end of the AFM tip, then we speak about single-molecule force spectroscopy (SMFS). The recorded force provides information on the binding strength between the grafted molecule and its receptors, the localisation of the latter, and the nanomechanics of the molecules which are being pulled.

In SMFS, the tip can be modified with chemical groups [23] or bioligands [24]. The precise movement of the sample and simultaneous measurement of the binding force allow to accurately map and follow the dynamic rearrangement of the receptors at the cell surface [25–29]. A mesh of force curves is therefore recorded across the cell surface, revealing the adhesion between the tip and the sample in every location. A two-dimensional map can be reconstructed where the brightness of the pixel reflects the magnitude of the force.

Recently, the technology has continued to evolve, enabling a single living microbe to replace the sharp tip of the cantilever. Interactions of the microorganism with a substrate can now be measured at the single-cell level. This development can be seen as the link between measurements done at the single-molecule scale and observations made by classical microbiological assays at the scale of the cell population.

1.3 Imaging the Surface of Living Pathogens

Given its very high resolution level combined with its ability to work in liquid conditions, AFM has already revealed fascinating features of the living cell surfaces. A key requirement to achieve proper imaging of live organisms is a good immobilization of the sample that will be analysed. When dealing with mammalian cells, one simple approach consists in allowing the cells to naturally spread and adhere to a flat surface [30] with the possibility to pre-coat the substrate with proteins to improve the cell anchorage [31]. However, streaks are often seen on the images arising from the tip scanning the soft cell membrane, leading sometimes to cell detachment. To overcome this issue, different protocols have been developed such as chemical fixation using cross-linking agents (mainly paraformaldehyde or glutaraldehyde) [25, 31, 32]. A recent method in which the concentration of glutaraldehyde is reduced to the minimum has been suggested to prevent osmotic changes and to let surface receptors active [25, 27]. Using this protocol, defined as 'gentle fixation', fine filamentous structures of 50–300 nm have been observed at the edge of macrophages, and have been attributed to 'filipodia', i.e. protrusions which contain parallel bundles of actin filaments [32].

Different strategies need to be adopted for proper immobilization of microbes due to their spherical and well-defined shapes. Various methods have therefore been proposed, such as coating the substrate with positively charged molecules (poly-L-lysine or poly-ethyleneimine) [33] or gelatin, although the contamination of the cells by these pre-coatings remain uncertain. Poly-dimethylsiloxane (PDMS) patterned microstructures have also been used for microbes deposition [34]. Probably the most elegant technique so far consists in mechanically capturing cells in the pores of filtering polymer membranes. By choosing a pore size diameter which matches the size of the microbes [35, 36] several cells can then be trapped in the support. This immobilization technique has allowed the direct observation of nanoscale changes occurring in microorganisms when subjected to drugs.

To fight *Candida* infections, three groups of drugs have been traditionally used: azoles, polyenes and pyrimidines [9, 12]. A new class of drugs, the echinocandins, has recently been developed to target the synthesis of the β-1,3-glucan of the *Candida* cell wall. Among these echinocandins is caspofungin, which can lead to cell lysis and death for *C. albicans* [37–40]. Using different modes of the AFM instrument, the effect of the drug on *C. albicans* cells was analyzed at the nanoscale [41] and correlated with previous electron microscopy data [37, 39]. Imaging caspofungin treated yeast revealed an altered morphology and increased cell roughness from $\sim$1 to $\sim$5 nm confirming that β-1,3-glucans serve as maintaining the cell shape (Fig. 1.1a, b; [41]). AFM tip was then used as a nanoindentor to probe the mechanical response of the treated (as compared to untreated) cell walls. *C. albicans* mechanical strength decreased upon drug injection, highlighting the role of the cell surface polysaccharides in the cell wall rigidity.

Morphological variation of the surface of cells subjected to environmental changes can also be visualized by AFM. A direct example is the observation of

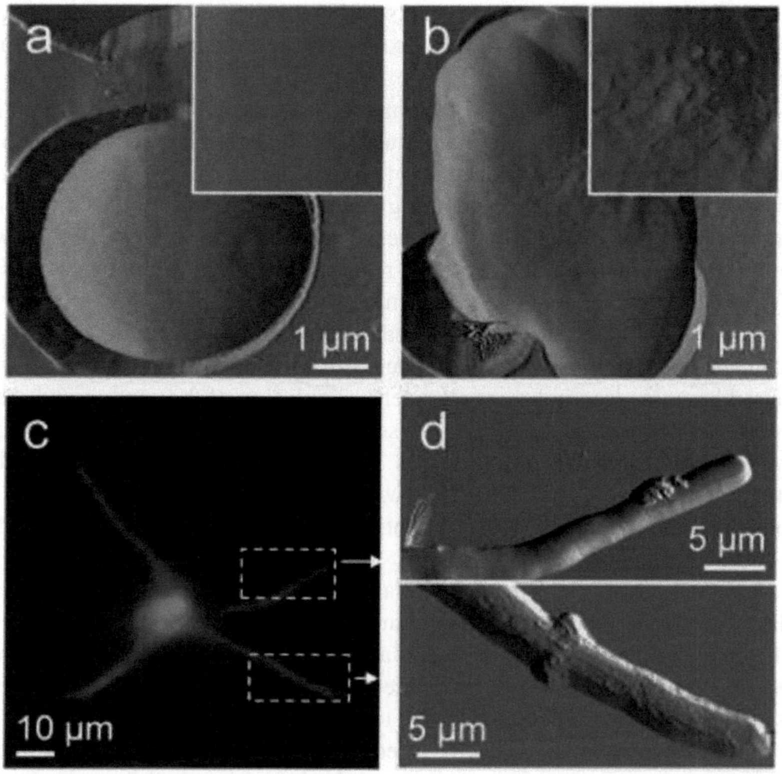

Fig. 1.1 AFM imaging captures *C. albicans* cell surface subjected to drugs and environmental stress. (**a, b**) Deflection images of native (**a**) and caspofungin-treated (**b**) *C. albicans* cells revealing the swelling and deformation of the cell upon drug injection. Insets show high resolution of the surface of the corresponding images (Adapted from [41] with permission from the Royal Society of Chemistry). (**c, d**) Fluorescence image (**c**) and correlative AFM images (**d**) of a macrophage (stained in *green*) incubated for 3 h with *C. albicans* cells (stained in *blue*). Images in (**d**) are defection images of the dashed area in (**c**) revealing major differences between free (*top*) and internalized (*bottom*) hyphae (Reprinted from [32] with permission from the American Chemical Society)

the surface dynamics of *A. fumigatus* during germination [42]. Understanding the cell wall properties of these aerial asexual spores is of primarily importance as they are the key players in the interactions with the host cells, which can lead to severe respiratory diseases [18, 43, 44]. Electron microscopy and biochemical studies had revealed that *A. fumigatus* conidia surface is covered by an assembly of proteinaceous microfibrils called rodlets [45, 46]. By using AFM to monitor the structural and physicochemical dynamics of the surface, Dague et al. have demonstrated how this clustered rodlets layer disappears as the conidia is subjected to germination process [42]. This capital transition is believed to be function-related [47]. Indeed, the spore external surface goes from an hydrophobic layer which promotes dispersion by air currents and mediates adhesion, to an amorphous hydrophilic material favouring hyphal growth through moist environments such as endothelia and epithelia [48].

As for fungal observations, AFM imaging has also offered the possibility to directly visualize the effect of antibacterial agents on bacterial surfaces. Antibiotics targeting *S. aureus*, such as β-lactams or glycopeptides, act by targeting enzymes involved in the synthesis of the peptidoglycan or by preventing the incorporation of peptidoglycan precursors into the bacterial cell wall, respectively [49]. However, given the constant increase of antibiotics-resistant strains, development of new anti-staphylococcal compounds is needed. Among such potential target is lysostaphin, an enzyme which cleaves the peptidoglycan of the bacteria. By recording time-lapse AFM images *of S. aureus* surface subjected to lysostaphin, Francius et al. nicely illustrated how the drug induces cell swelling, splitting of the septum and creation of nanoscale perforation [50]. They also revealed the changes in the bacterial cell wall nanomechanical properties as lysostaphin digested the peptidoglycan, resulting in osmotically fragile bacteria with reduced cell wall stiffness. This study demonstrated the lytic activity of new anti-staphylococcal agents and showed their great potential in nanomedicine.

More recent advances in the AFM platforms have also enabled a better understanding of the interactions between fungal pathogens and our immune system. One key step in developing this understanding has been the development of set-ups where AFM and correlative fluorescence imaging can be done simultaneously on living cells. Taking advantage of the technique, we have captured the different phases occurring when *C. albicans* infect macrophages [32]. We visualized the main steps of the interaction, starting from the initial intercellular contact, to the internalization of the pathogen followed by the germination where the hyphal growth leads to the macrophage membrane extension (Fig. 1.1c, d). Fluorescence gave the ability to clearly locate the pathogens even once internalized by the macrophages; whereas AFM revealed the macrophage surface structures with fine details, which were highly biologically relevant to a better understanding of the interaction [32].

1.4 Localizing and Manipulating Pathogens Surface Components

Adhesion is the crucial initial step of microbial infections and biofilm formation. Understanding the role of the molecules involved in this key process is necessary when one wants to control pathogens attachment and develop new therapeutic agents. While imaging mode offers unprecedent resolution of the nano-architecture of cells, it lacks in biochemical specificity. To overtake this drawback, AFM tips are modified with specific bioligands, making it possible to map their surface associated receptors.

Such single-molecule experiments have allowed to decipher a variety of mechanims responsible for the adhesion of *C. albicans*. For this species, the adhesion is known to be mainly mediated by cell wall-anchored glycoproteins, so called Als adhesins [14, 51, 52]. Eight different genes sharing common sequences encode this family of adhesins. As a result, Als proteins present similar compositions, with an N terminal Ig-like region that initiates cell adhesion, a threonine rich region with amyloid sequence, a tandem repeat region for cell-cell aggregation, and a stalk on C terminal region projecting the adhesive part away from the cell surface [51, 52].

By combining SMFS with genetic manipulation, Alsteens et al. have demonstrated that if one protein of this superfamily, Als5, is expressed in non-pathogenic yeast species, its nanomechanical response under force could be clearly identified [26, 53]. Using Als5 proteins modified with a V5 epitope tag [53] and anti-V5 AFM tips, the unfolding of the Als proteins submitted to pulling force was unraveled, reminiscent of the unfolding of the mammalian extracellular matrix protein tenascin [54] or of the titin [55]. Force curves of unfolding events presented a sawtooth pattern in which each peak has been attributed to the unfolding of an individual tandem repeat (Fig. 1.2b). As Als tandem repeats are known to contain hydrophobic aminoacids [56], the exposure of these features when the proteins are in an extended conformation was hypothesized to favor hydrophobic interactions between cells.

Additionally, these studies have permitted the following of dynamic reorganization of Als5 proteins when subjected to force. The adhesins were seen to form clusters resulting in adhesive nanodomains, which contribute to firm attachment of the microorganism to surfaces (Fig. 1.2c, d [26]). The concept of nanodomains has also been observed for the distribution of cadherins at the surface of vascular endothelial cells [27], mycobacterial adhesions involved in host interaction [28], or staphylococcal adhesions [21] suggesting that force-induced clustering could be a general mechanism for pathogens adhesion.

SMFS has also been used to understand the effect of caspofungin on *C. albicans* cell wall composition [41]. At sub inhibitory concentration, the drug induces formation of cell aggregates resulting from overexpression of another Als adhesion, namely Als1 [41, 57]. Using tips functionalized with a peptide recognized by Als1 proteins, we showed how caspofungin treatment can induce molecular remodeling of *C. albicans* cell wall, where Als1 overexposure promotes cell aggregation (Fig. 1.2e, f; [41]). These experiments support the proposed mechanism in which the tandem repeats of extended Als would generate an enhanced hydrophobicity triggering cell-cell aggregation as a stress response.

Recently, we used SMFS to decipher the nanoscale modification of surface composition during the yeast-to-hyphae transition occurring for *C. albicans*, also called morphogenesis. This strong virulence factor has been demonstrated to be concomitant of molecular cell wall changes, as demonstrated by fluorescence and western blot studies [3, 14, 58–60]. Indeed, germ tube formation has been linked to an overexpression of Als3 proteins in the cell wall [61]. Using tips functionalized with anti-Als3 antibodies, we observed that the exposure and nanomechanical properties of the adhesins drastically change upon germination (Fig. 1.2g, h; [62]). Als3 were found to be much more present on the germ tube as compared to the

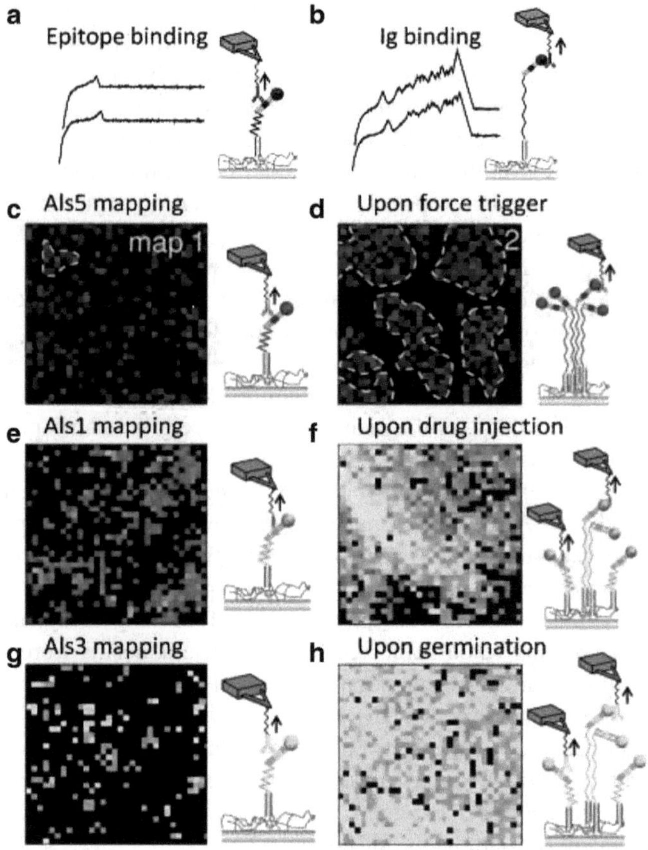

Fig. 1.2 Mapping of the major *C. albicans* adhesins upon environmental changes. (**a**, **b**) Force curves representative of the single weak epitope binding (**a**) or featured sawtooth patterns with multiple force peaks arising from the unfolding of the Als proteins upon immunoglobulin binding to the functionalized AFM tip (**b**). (**c**–**h**) Adhesion force maps taken at the cell surface mapping the distribution of Als5 proteins (**c**) clustering to form adhesive nanodomains upon force trigger (**d**); distribution of Als1 proteins (**e**) and the overexpression of the proteins upon capsofungin injection (**f**) and distribution of the Als3 proteins at the surface of the budding yeast (**g**) and the hyphae (**h**) of *C. albicans* (Maps in (**c**) and (**d**) have been reprinted from [26] with permission from the National Academy of Science; Curves in (**a**, **b**) and maps in (**g**, **h**) have been modified from [62] with permission from the American Chemical Society; maps in (**e**, **f**) have been reprinted from [41] with permission of the Royal Society of Chemistry)

budding yeast leading to an increased adhesion of hyphae as compared to yeast or mother cell. This transition was also accompanied by an increase hydrophobicity and increased exposure and extension of the surface mannan glycoconjugates. All together, these findings suggest that germination promotes adhesion to various types of surfaces, including hydrophobic implants (such as catheters), epithelial and endothelial mammalian cells decorated with Als3 receptors, but also to immune cells covered with mannose receptors.

Single-molecule experiments have also answered a number of questions on the adhesion of pathogenic bacteria. *Burkholderia cenocepacia* is a bacterium responsible for severe infections in patients with cystic fibrosis [63–65]. SMFS has contributed to elucidate the role of the trimeric autotransporters TAAs in the binding mechanisms of *B. cenocepacia* [66]. The proteins were shown to form homophilic trans-interactions involved in bacterial aggregation. They are also capable to withstand very high forces by acting as a nanospring, a mechanism reminiscent of pili decorating the gram-positive probiotic bacteria *Lactobacillus rhamnosus* GG [67]. TAAs were also demonstrated to bind to host epithelial cells, either via a major extracellular component, collagen, or by recognizing specific receptors on living pneumocytes leading to the formation of membrane tethers. Finally, the peculiar strength of the attachment force of the *S. epidermidis* to fibrinogen has been recently discovered using SMFS [21]. Single-cell surface adhesins, SdrG, were seen to attach to fibrinogen coated tips with a force of 2 nN, equivalent to a covalent bond, and to cluster in super-adhesive nanodomains [21]. These mechanisms would help *S. epidermidis* to colonize implanted biomaterials and to withstand physiological shear forces. All together, these findings shed new light on the adhesive and mechanical properties of pathogenic bacteria and fungi.

1.5 Quantifying the Force Driving Single Cell

We have demonstrated the ability of SMFS to decipher the role played by individual components at the cell surface. However, adhesion usually involves an interplay of physico-chemical forces, including specific receptor-ligand forces and non-specific hydrophobic and electrostatic forces [68, 69]. Therefore, the need to measure the global force guiding cell attachment to another cell or to solid substrate rapidly appeared. In that sense, single-cell force spectroscopy was first developed for mammalian cells [70–73]. Since then, SCFS has proven to be instrumental in unraveling the fundamental forces driving cell-substrate and cell-cell adhesion.

In a same context, SCFS has been extended to microbes. Using wedged tipless cantilevers [74] coated with a thin film of polydopamine [75], a single *C. albicans* yeast could be immobilized on the cantilever [76, 77]. The development of *C. albicans* biofilm starts with the formation of an initial based layer composed of budding yeast cells adhering to the substrate. Then the upper layer develops by formation of microcolonies and growing of hyphaes [3]. Therefore, Alsteens et al. have used SCFS of *C. albicans* yeast towards germinated cells to get a better indication of the contribution of the different fungal morphotypes within the biofilm. They proposed a model where strong cohesive adhesion forces between yeast and hyphae would originate from tight hydrophobic interactions between Als tandem repeats domains on adjacent cells, i.e. Als3 proteins on the germ tubes and other Als proteins such as Als1 or Als5 on the yeasts [77].

Polymicrobial infections generally involve the formation of mixed biofilm, where fungal and bacterial pathogens co-adhere to each other [78–81]. An example of such

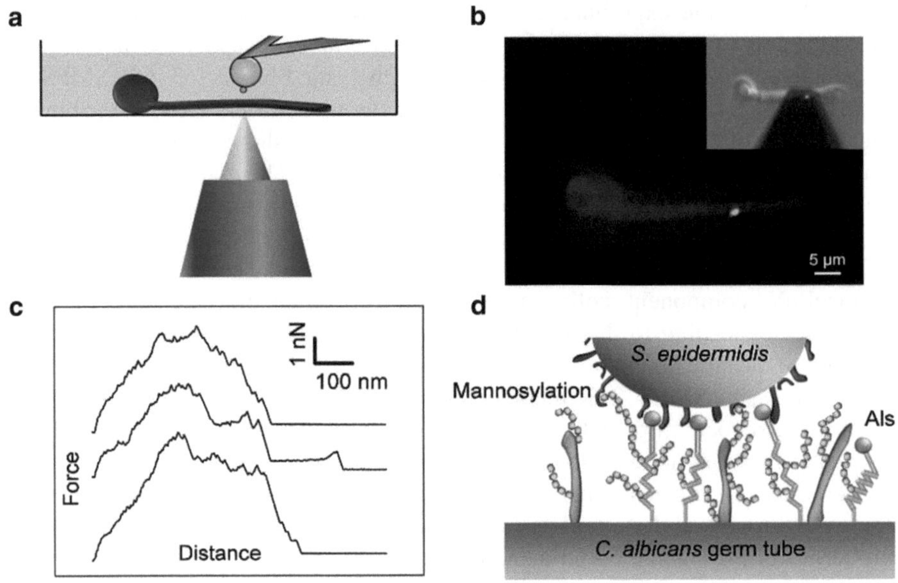

Fig. 1.3 Quantification of the specific forces engaged in bacterial-fungal interactions. (**a**) Schematic of the set-up. (**b**) Using an inverted fluorescence microscope combined to an AFM, *S. epidermidis* (stained in *green*) is approached toward *C. albicans* (stained in *blue*). (**c**) Typical force-distance curves recorded in buffer between the bacterium and the *C. albicans* hyphae; (**d**) Main *C. albicans* cell wall components involved in the interactions are surface glycoproteins (*green*) and mannose-rich glycoconjugates (*blue*) (Adapted from [87] with permission from the Royal Society of Chemistry)

association is *C. albicans* and *S. aureus*, whose synergistic effect has been shown to lead to mortality increase [82–84]. Also of primarily interest are the mixed species biofilm composed of the two nosocomial pathogens *C. albicans* and *S. epidermidis*. To quantify the forces involved in such biofilms, we first developed a reliable and simple method to attach a single bacterium to the AFM cantilever (Fig. 1.3a, b). For that, we use colloidal probes coated with bioinspired polydopamine wet adhesive [85, 86]. With this set-up, proper position and viability of the attached bacteria was ensured (Fig. 1.3b). We then recorded force curves between such bacterial probes and *C. albicans* germ tubes, where the adhesion was found to be as strong as 5 nN (Fig. 1.3c; [87]). By contrast, *S. epidermidis* adhere to a lower extend on yeast cells, emphasizing the role played by the *C. albicans* morphological transition in mediating adhesion to bacterial cells. We further elucidated the main fungal surface components involved in this strong adhesion (mainly Als proteins and O-mannosylations, Fig. 1.3d) by using *C. albicans* mutants altered in their cell wall composition [87].

Finally, Herman et al. also used this approach to attach single *S. epidermidis* WT and mutated cells to the cantilever to unravel how SdrG mediates time-dependent attachment to fibrinogen-coated substrates [21].

1.6 Conclusions

AFM is a powerful versatile platform to understand the main mechanisms responsible of pathogens adhesion. This nanotechnology has already given breakthrough insights onto the molecular origin of pathogens attachment to substrates, to other microorganisms and to mammalian cells. It has offered exiting avenues that help design molecules capable of inhibiting biofilm formation.

Although already largely employed, the AFM toolbox is still growing with the development of new methodologies aiming to image microbes and probe interaction forces with unprecedented speed and resolution. In that direction, advanced multiparametric AFM technology combined with biochemically tips has recently permitted to locate bacteriophages extruding from the surface of *Escherichia coli* and to relate their position to the nanomechanical properties of the bacterial cell wall [88]. Using high-speed AFM, the dynamic effect of antimicrobial peptides on *E. coli* has been decomposed in a combination of incubation and execution phases proving the enormous potential of this technique to visualize cellular processes in real time and at the nanoscale [89]. Finally the diversity and endless possibility of AFM applications has been demonstrated recently, where fluctuations of highly sensitive AFM cantilevers were recorded as a sensor to characterize the metabolism of *E. coli* and *S. aureus* under the effect of antibiotics [90]. We anticipate that AFM measurements will be increasingly used to decipher pathogens biofilm formation and developing new therapeutic agents.

References

1. Costerton JW, Stewart PS, Greenberg EP (1999) Bacterial biofilms: a common cause of persistent infections. Science 284:1318–1322
2. Douglas LJ (2003) *Candida* biofilms and their role in infection. Trends Microbiol 11:30–36
3. Finkel JS, Mitchell AP (2011) Genetic control of *Candida albicans* biofilm development. Nat Rev Microbiol 9:109–118
4. Foster TJ, Geoghegan JA, Ganesh VK et al (2014) Adhesion, invasion and evasion: the many functions of the surface proteins of *staphylococcus aureus*. Nat Rev Microbiol 12:49–62
5. Kolter R, Greenberg EP (2006) Microbial sciences: the superficial life of microbes. Nature 441:300–302
6. Otto M (2009) *Staphylococcus epidermidis* – the 'accidental' pathogen. Nat Rev Microbiol 7:555–567
7. Uckay I, Pittet D, Vaudaux P et al (2009) Foreign body infections due to *Staphylococcus epidermidis*. Ann Med 41:109–119
8. Pfaller MA, Diekema DJ (2007) Epidemiology of invasive candidiasis: a persistent public health problem. Clin Microbiol Rev 20:133–163
9. Pfaller MA, Espinel-Ingroff A, Canton E et al (2012) Wild-type MIC distributions and epidemiological cutoff values for amphotericin B, flucytosine, and itraconazole and Candida spp. as determined by CLSI broth microdilution. J Clin Microbiol 50:2040–2046
10. Roetzer A, Gabaldon T, Schuller C (2011) From *Saccharomyces cerevisiae* to *Candida glabrata* in a few easy steps: important adaptations for an opportunistic pathogen. FEMS Microbiol Lett 314:1–9

11. Calderone R (2004) The 7th conference on *Candida* and Candidiasis, Austin, Texas, USA, March 18–22, 2004. FEMS Yeast Res 4:885–886
12. Odds FC, Brown AJ, Gow NA (2003) Antifungal agents: mechanisms of action. Trends Microbiol 11:272–279
13. Pfaller MA (2012) Antifungal drug resistance: mechanisms, epidemiology, and consequences for treatment. Am J Med 125:S3–S13
14. Gow NAR, van de Veerdonk FL, Brown AJP et al (2012) *Candida albicans* morphogenesis and host defense: discriminating invasion from colonization. Nat Rev Microbiol 10:112–122
15. Netea MG, Gow NA, Munro CA et al (2006) Immune sensing of *Candida albicans* requires cooperative recognition of mannans and glucans by lectin and Toll-like receptors. J Clin Invest 116:1642–1650
16. Netea MG, Brown GD, Kullberg BJ et al (2008) An integrated model of the recognition of *Candida albicans* by the innate immune system. Nat Rev Microbiol 6:67–78
17. Mahdavinia M, Grammer LC (2012) Management of allergic bronchopulmonary aspergillosis: a review and update. Ther Adv Respir Dis 6:173–187
18. Beauvais A, Fontaine T, Aimanianda V et al (2014) *Aspergillus* cell wall and biofilm. Mycopathologia 178:371–377
19. Aimanianda V, Bayry J, Bozza S et al (2009) Surface hydrophobin prevents immune recognition of airborne fungal spores. Nature 460:1117–1121
20. Fontaine T, Delangle A, Simenel C et al (2011) Galactosaminogalactan, a new immunosuppressive polysaccharide of *Aspergillus fumigatus*. PLoS Pathog 7:e1002372
21. Herman P, El-Kirat-Chatel S, Beaussart A et al (2014) The binding force of the staphylococcal adhesin SdrG is remarkably strong. Mol Microbiol 93:356–368
22. Foster TJ (2005) Immune evasion by staphylococci. Nat Rev Microbiol 3:948–958
23. Frisbie CD, Rozsnyai LF, Noy A et al (1994) Functional group imaging by chemical force microscopy. Science 265:2071–2074
24. Hinterdorfer P, Dufrene YF (2006) Detection and localization of single molecular recognition events using atomic force microscopy. Nat Methods 3:347–355
25. Ahmad SF, Chtcheglova LA, Mayer B et al (2011) Nanosensing of Fcgamma receptors on macrophages. Anal Bioanal Chem 399:2359–2367
26. Alsteens D, Garcia MC, Lipke PN et al (2010) Force-induced formation and propagation of adhesion nanodomains in living fungal cells. Proc Natl Acad Sci U S A 107:20744–20749
27. Chtcheglova LA, Waschke J, Wildling L et al (2007) Nano-scale dynamic recognition imaging on vascular endothelial cells. Biophys J 93:L11–L13
28. Dupres V, Alsteens D, Wilk S et al (2009) The yeast Wsc1 cell surface sensor behaves like a nanospring in vivo. Nat Chem Biol 5:857–862
29. Kim H, Arakawa H, Hatae N et al (2006) Quantification of the number of EP3 receptors on a living CHO cell surface by the AFM. Ultramicroscopy 106:652–662
30. Radmacher M, Tillamnn RW, Fritz M et al (1992) From molecules to cells: imaging soft samples with the atomic force microscope. Science 257:1900–1905
31. Labernadie A, Thibault C, Vieu C et al (2010) Dynamics of podosome stiffness revealed by atomic force microscopy. Proc Natl Acad Sci U S A 107:21016–21021
32. El-Kirat-Chatel S, Dufrene YF (2012) Nanoscale imaging of the *Candida*-macrophage interaction using correlated fluorescence-atomic force microscopy. ACS Nano 6:10792–10799
33. Schaer-Zammaretti P, Ubbink J (2003) Imaging of lactic acid bacteria with AFM-elasticity and adhesion maps and their relationship to biological and structural data. Ultramicroscopy 97:199–208
34. Dague E, Jauvert E, Laplatine L et al (2011) Assembly of live micro-organisms on microstructured PDMS stamps by convective/capillary deposition for AFM bio-experiments. Nanotechnology 22:395102
35. Dufrêne YF (2008) Atomic force microscopy and chemical force microscopy of microbial cells. Nat Protoc 3:1132–1138
36. Kasas S, Ikai A (1995) A method for anchoring round shaped cells for atomic force microscope imaging. Biophys J 68:1678–1680

37. Bizerra FC, Melo AS, Katchburian E et al (2011) Changes in cell wall synthesis and ultrastructure during paradoxical growth effect of caspofungin on four different *Candida* species. Antimicrob Agents Chemother 55:302–310
38. Chaffin WL, Lopez-Ribot JL, Casanova M et al (1998) Cell wall and secreted proteins of *Candida albicans*: identification, function, and expression. Microbiol Mol Biol Rev 62:130–180
39. Dunyach C, Drakulovski P, Bertout S et al (2011) Fungicidal activity and morphological alterations of *Candida albicans* induced by echinocandins: study of strains with reduced caspofungin susceptibility. Mycoses 54:e62–e68
40. Letscher-Bru V, Herbrecht R (2003) Caspofungin: the first representative of a new antifungal class. J Antimicrob Chemother 51:513–521
41. El-Kirat-Chatel S, Beaussart A, Alsteens D et al (2013) Nanoscale analysis of caspofungin-induced cell surface remodelling in *Candida albicans*. Nanoscale 5:1105–1115
42. Dague E, Alsteens D, Latgé JP et al (2008) High-resolution cell surface dynamics of germinating *Aspergillus fumigatus* conidia. Biophys J 94:656–660
43. Annaix V, Bouchara JP, Larcher G et al (1992) Specific binding of human fibrinogen fragment D to *Aspergillus fumigatus* conidia. Infect Immun 60:1747–1755
44. Sheppard DC (2011) Molecular mechanism of *Aspergillus fumigatus* adherence to host constituents. Curr Opin Microbiol 14:375–379
45. Beever RE, Dempsey GP (1978) Function of rodlets on the surface of fungal spores. Nature 272:608–661O
46. Cole GT, Sekiya T, Kasai R et al (1979) Surface ultrastructure and chemical composition of the cell walls of conidial fungi. Exp Mycol 3:132–156
47. Wessels JGH (1996) Fungal hydrophobins: proteins that function at an interface. Trends Plant Sci 1:9–15
48. Filler SG, Sheppard DC (2006) Fungal invasion of normally non-phagocytic host cells. PLoS Pathog 2:1099–1105
49. Walsh C (2000) Molecular mechanisms that confer antibacterial drug resistance. Nature 406:775–781
50. Francius G, Domenech O, Mingeot-Leclercq MP et al (2008) Direct observation of *Staphylococcus aureus* cell wall digestion by lysostaphin. J Bacteriol 190:7904–7909
51. Hoyer LL (2001) The ALS gene family of *Candida albicans*. Trends Microbiol 9:176–180
52. Lipke PN, Garcia MC, Alsteens D et al (2012) Strengthening relationships: amyloids create adhesion nanodomains in yeasts. Trends Microbiol 20:59–65
53. Alsteens D, Dupres V, Klotz SA et al (2009) Unfolding individual Als5p adhesion proteins on live cells. ACS Nano 3:1677–1682
54. Oberhauser AF, Marszalek PE, Erickson HP et al (1998) The molecular elasticity of the extracellular matrix protein tenascin. Nature 393:181–185
55. Rief M, Gautel M, Oesterhelt F et al (1997) Reversible unfolding of individual titin immunoglobulin domains by AFM. Science 276:1109–1112
56. Dranginis AM, Rauceo JM, Coronado JE et al (2007) A biochemical guide to yeast adhesins: glycoproteins for social and antisocial occasions. Microbiol Mol Biol Rev 71:282–294
57. Gregori C, Glaser W, Frohner IE et al (2011) Efg1 controls caspofungin-induced cell aggregation of *Candida albicans* through the adhesin Als1. Eukaryot Cell 10:1694–1704
58. Hazen BW, Hazen KC (1988) Dynamic expression of cell surface hydrophobicity during initial yeast cell growth and before germ tube formation of *Candida albicans*. Infect Immun 56:2521–2525
59. Liu Y, Filler SG (2011) *Candida albicans* Als3, a multifunctional adhesin and invasin. Eukaryot Cell 10:168–173
60. Nobile CJ, Schneider HA, Nett JE et al (2008) Complementary adhesin function in *C. albicans* biofilm formation. Curr Biol 18:1017–1024
61. Coleman DA, Oh SH, Zhao X et al (2009) Monoclonal antibodies specific for *Candida albicans* Als3 that immunolabel fungal cells in vitro and in vivo and block adhesion to host surfaces. J Microbiol Methods 78:71–78

62. Beaussart A, Alsteens D, El-Kirat-Chatel S et al (2012) Single-molecule imaging and functional analysis of Als adhesins and mannans during *Candida albicans* morphogenesis. ACS Nano 6:10950–10964
63. Mil-Homens D, Fialho AM (2011) Trimeric autotransporter adhesins in members of the *Burkholderia cepacia* complex: a multifunctional family of proteins implicated in virulence. Front Cell Infect Microbiol 1:13
64. Mil-Homens D, Fialho AM (2012) A BCAM0223 mutant of *Burkholderia cenocepacia* is deficient in hemagglutination, serum resistance, adhesion to epithelial cells and virulence. PLoS One 7:e41747
65. Mil-Homens D, Rocha EPC, Fialho AM (2010) Genome-wide analysis of DNA repeats in *Burkholderia cenocepacia* J2315 identifies a novel adhesin-like gene unique to epidemic-associated strains of the ET-12 lineage. Microbiology 156:1084–1096
66. El-Kirat-Chatel S, Mil-Homens D, Beaussart A et al (2013) Single-molecule atomic force microscopy unravels the binding mechanism of a *Burkholderia cenocepacia* trimeric autotransporter adhesin. Mol Microbiol 89:649–659
67. Tripathi P, Beaussart A, Alsteens D et al (2013) Adhesion and nanomechanics of pili from the probiotic *Lactobacillus rhamnosus* GG. ACS Nano 7:3685–3697
68. Busscher HJ, Norde W, Van Der Mei HC (2008) Specific molecular recognition and nonspecific contributions to bacterial interaction forces. Appl Environ Microbiol 74:2559–2564
69. Busscher HJ, van der Mei HC (2012) How do bacteria know they are on a surface and regulate their response to an adhering state? PLoS Pathog 8:e1002440
70. Benoit M, Gabriel D, Gerisch G et al (2000) Discrete interactions in cell adhesion measured by single-molecule force spectroscopy. Nat Cell Biol 2:313–317
71. Benoit M, Gaub HE (2002) Measuring cell adhesion forces with the atomic force microscope at the molecular level. Cells Tissues Organs 172:174–189
72. Friedrichs J, Helenius J, Muller DJ (2010) Quantifying cellular adhesion to extracellular matrix components by single-cell force spectroscopy. Nat Protoc 5:1353–1361
73. Helenius J, Heisenberg CP, Gaub HE et al (2008) Single-cell force spectroscopy. J Cell Sci 121:1785–1791
74. Stewart MP, Hodel AW, Spielhofer A et al (2013) Wedged AFM-cantilevers for parallel plate cell mechanics. Methods 60:186–194
75. Lee H, Dellatore SM, Miller WM et al (2007) Mussel-inspired surface chemistry for multifunctional coatings. Science 318:426–430
76. Alsteens D, Beaussart A, Derclaye S et al (2013) Single-cell force spectroscopy of Als-mediated fungal adhesion. Anal Methods 5:3657–3662
77. Alsteens D, Van Dijck P, Lipke PN et al (2013) Quantifying the forces driving cell-cell adhesion in a fungal pathogen. Langmuir 29:13473–13480
78. Elias S, Banin E (2012) Multi-species biofilms: living with friendly neighbors. FEMS Microbiol Rev 36:990–1004
79. Harriott MM, Noverr MC (2009) *Candida albicans* and *Staphylococcus aureus* form polymicrobial biofilms: effects on antimicrobial resistance. Antimicrob Agents Chemother 53:3914–3922
80. Hogan DA, Kolter R (2002) *Pseudomonas-Candida* interactions: an ecological role for virulence factors. Science 296:2229–2232
81. Lynch AS, Robertson GT (2008) Bacterial and fungal biofilm infections. Annu Rev Med 59:415–428
82. Carlson E (1983) Effect of strain of *Staphylococcus aureus* on synergism with *Candida albicans* resulting in mouse mortality and morbidity. Infect Immun 42:285–292
83. Morales DK, Hogan DA (2010) *Candida albicans* interactions with bacteria in the context of human health and disease. PLoS Pathog 6:1–4
84. Shirtliff ME, Peters BM, Jabra-Rizk MA (2009) Cross-kingdom interactions: *Candida albicans* and bacteria. FEMS Microbiol Lett 299:1–8
85. Beaussart A, El-Kirat-Chatel S, Herman P et al (2013) Single-cell force spectroscopy of probiotic bacteria. Biophys J 104:1886–1892

86. Beaussart A, El-Kirat-Chatel S, Sullan RMA et al (2014) Quantifying the forces guiding microbial cell adhesion using single-cell force spectroscopy. Nat Protoc 9:1049–1055
87. Beaussart A, Herman P, El-Kirat-Chatel S et al (2013) Single-cell force spectroscopy of the medically important *Staphylococcus epidermidis-Candida albicans* interaction. Nanoscale 5:10894–10900
88. Alsteens D, Trabelsi H, Soumillion P et al (2013) Multiparametric atomic force microscopy imaging of single bacteriophages extruding from living bacteria. Nat Commun 4:1–7
89. Fantner GE, Barbero RJ, Gray DS et al (2010) Kinetics of antimicrobial peptide activity measured on individual bacterial cells using high-speed atomic force microscopy. Nat Nanotechnol 5:280–285
90. Longo G, Alonso-Sarduy L, Rio LM et al (2013) Rapid detection of bacterial resistance to antibiotics using AFM cantilevers as nanomechanical sensors. Nat Nanotechnol 8:522–526

Chapter 2
Bacteriophage-based Products and Techniques for Identification of Biological Pathogens

Nina Chanishvili

Abstract Bacteriophage (phage) are obligate intracellular parasites that multiply inside bacteria by making use of some or all of the host biosynthetic machinery (i.e., viruses that infect bacteria). Bacteriophages are highly specific to their host pathogens.

An idea of using phages against bacterial infections, known as phage therapy, appeared practically simultaneously with the discovery of phages. However, after discovery of antibiotics, the interest of Western clinicians turned to use of broad-spectrum antibiotics for treating unknown infections that kill many types of bacterium. Phages, by contrast, kill just one species or strain. With the rise of antibiotic-resistance the researchers now realize that they need more precise ways to target multi-drug-resistant pathogenic bacteria to which bacteriophages are still active.

Recently, the nanosized dimensions of bacteriophages, coupled with the ease of use with which genetic modifications can be made to their structure and function, have put phages in the spotlight. This is because of their use in a variety of biosensors, including as recognition probes for pathogen detection. The popular bio-probes that have been employed on biosensor surface for pathogen detection are nucleic acids, antibodies, whole phages, phage-display peptides (PDPs), and most recently phage's receptor binding proteins (RBPs). The efforts are now directed to development of enhanced detection technologies with high levels of reliability, sensitivity, and selectivity with short assay times.

This review summarizes experiences in development of phage-based therapeutics, prophylactic and diagnostic preparations, and their uses in different fields, such as medicine, veterinary, agriculture, food and water safety, etc.

N. Chanishvili (✉)
Eliava Institute of Bacteriophage, Microbiology and Virology, Gotua street 3, 0160 Tbilisi, Georgia
e-mail: nina.chanishvili@gmail.com

T.A. Camesano (ed.), *Nanotechnology to Aid Chemical and Biological Defense*, NATO Science for Peace and Security Series A: Chemistry and Biology,
DOI 10.1007/978-94-017-7218-1_2

2.1 Bacteriophages

Bacteriophage, or phage for short, are viruses that infect only bacteria. In contrast to cells that grow from an increase in the number of their components and reproduce by division, viruses are assembled from pre-made components. Viruses are nucleic acid molecules surrounded by a protective coating. They are not capable of generating energy and reproduce inside of cells. The nucleic acid inside the coating, called the phage genome in a bacteriophage, encodes most of the gene products needed for making more phage. The phage genome can be made of either double- or single-stranded DNA or RNA, depending on the bacteriophage in question. The genome can be circular or linear. The protective coating or capsid surrounding the phage genome is composed of phage-encoded proteins.

Many important discoveries have been made using phage as model systems. From the discovery that a nonsense codon stopped protein synthesis, to the first developmental switch to be understood at the molecular level, phage have proven to be very useful.

2.2 Discovery of Bacteriophages

During the First World War a rather sensational news was spread out – the viruses "eaters of micro-organisms" had been discovered by Felix d'Herelle, who developed a phage preparation to treat World War I soldiers with dysentery [29]. Doctors all over the world were excited with this news, especially because of the announcement published in medical journals that bacteriophages are harmless for both humans and animals, and can be successfully applied as therapeutic means [2, 8, 14, 15, 56, 61, 67].

Before that, in 1915, a well-known British journal "The Lancet" published an article written by Frederick Twort about "the transmissible bacterial lyses" [73], in which Twort described his observation of "the eaten edges of the colonies of Staphylococcus". He managed to filter the appropriate cultures of *Staphylococcus* and spotted the filtrate on the lawn of different *Staphylococcus* strains. Thus, he received a clear zone of lysis again and again. However, Twort could not explain the observed event and provided only its description. This was the very first publication on bacteriophages. French Canadian scientists Felix d'Herelle read this article and recalled in his mind his own observations in Mexico and Tunis [69]. He suspected that the filtered agent is a bacterial virus, an invisible invader that destroys bacteria.

The discovery, or rediscovery, of bacteriophages by Felix d'Herelle after Frederick Twort is frequently associated with an outbreak of severe hemorrhagic dysentery among French troops stationed at Maisons-Laffitte (on the outskirts of Paris) in July–August 1915 [67, 69]. Several soldiers were hospitalized, and d'Herelle was assigned to conduct an investigation of the outbreak. During these studies, he made bacterium-free filtrates of the patients' fecal samples, and mixed and incubated them with *Shigella* strains isolated from the patients. A portion of the mixtures was inoculated into experimental animals (as part of d'Herelle's studies on developing a vaccine against bacterial dysentery), and a portion was spread on agar medium in order to observe the growth of the bacteria. On these agar cultures, d'Herelle observed again the appearance of small, clear areas, which he initially called *taches*, then *taches vierges* (virgin spots), and, later, *plaques* [68]. D'Herelle's findings were presented during the September 1917 meeting of the Academy of Sciences, and they were subsequently published in the meeting's proceedings [31]. In contrast to Twort, d'Herelle had little doubt about the nature of the phenomenon, and he proposed that it was caused by a virus capable of parasitizing bacteria. The name "bacteriophage" (from "bacteria" and Greek φᾰγεῖν *phagein* "to eat") was also proposed by d'Herelle [68].

The discovery of bacteriophages was inevitable. Similar phenomena had been observed in remote regions of the world by different scientists. At the end of the nineteenth century, N.F. Gamaleya (later Honorary Member of the Academy of the USSR) published an article in the Russian Archives of the Pathological and Clinical Medicine [20]. In this article he described the lysis of *Bacillus antracis* in distilled water, after which the water obtained an ability to lyse the other strains of *Bacillus antracis*. In 1917, a young Georgian scientist George Eliava had observed mysterious disappearance of *Vibrio cholerae* cells [22, 30].

The greatest merit of Felix d'Herelle is that he advanced an idea of using bacteriophages for treatment of human and animal bacterial diseases. For this idea nowadays he would deserve the Noble Prize to which he was nominated eight times since 1925, though he was never awarded one (cited by Hausler [29] according to Nobel Archives).

2.3 Early Clinical Trials and Commercial Production of Bacteriophages

Not long after his discovery, d'Herelle used phages to treat dysentery in what was probably the first attempt to use bacteriophages therapeutically. The studies were conducted at the Hôpital des Enfants-Malades in Paris in 1919 under the clinical supervision of Professor Victor-Henri Hutinel, the hospital's Chief of Pediatrics [68]. The phage preparation was ingested by d'Herelle, Hutinel, and several hospital interns in order to confirm its safety before administering it the next day to a 12-year-old boy with severe dysentery. The patient's symptoms ceased after a single administration of d'Herelle's anti-dysentery phage, and the boy fully recovered within a few days. The efficacy of the phage preparation was "confirmed" shortly afterwards, when three additional patients having bacterial dysentery and treated with one dose of the preparation started to recover within 24 h of treatment. However, the results of these studies were not immediately published and, therefore, the first reported application of phages to treat infectious diseases of humans came later from Richard Bruynoghe and Joseph Maisin [8], who used bacteriophages to treat staphylococcal skin disease. The bacteriophages were injected into and around surgically opened lesions, and the authors reported regression of the infections within 24–48 h. Several similarly promising studies followed [61, 63, 66]. Encouraged by these early results, d'Herelle and others continued studies of the therapeutic use of phages; for example, d'Herelle used various phage preparations to treat thousands of people having cholera and/or bubonic plague in India [37, 68, 69].

In 1916–1930 d'Herelle and his collaborators undertook numerous expeditions to China, Laos, India, Vietnam, and Africa to combat epidemics caused by cholera and plague with bacteriophages. According to Romanian medical historians Kazhal and Iftimovich [37] the first attempts to use cholera bacteriophage for treatment and prophylaxis were performed by Felix d'Herelle and George Eliava in 1931. According to these authors, the Institute of Vaccine and Sera in Tbilisi (nowadays known as the Eliava Institute of Bacteriophage, Microbiology and Virology) produced the first commercial anticholera phage preparation, which was reported to be successfully used for control of epidemics threatening the South-East territories of the USSR [37]. According to the estimations that were published that time, the application of bacteriophages caused the mortality of cholera in India to be reduced to 10 % [37]. This fact is described by d'Herelle himself in the book "Bacteriophage and the Phenomenon of Recovery", published in the Russian language in 1935 in Tbilisi, Georgia.[1]

According to d'Herelle [30], cholera epidemics occurred in Punjab region in 1927. Patients were treated orally with 2 ml of cholera phage diluted in 20 ml of water but the titer of phage in these preparations is unknown. If the patient vomited, a repeated dose of phage (5 ml diluted in 100 ml of water) was administered slowly with a teaspoon. The control group of patients treated themselves using a folk medicine (plant extracts). Of 14,450 people who lived in nine villages of Punjab region, only 73 have been treated with the phage. D'Herelle explained this by saying that people in India opposed any new medical measures, and rarely permitted him and his colleagues to use phages for treatment. Thus, only desperately ill patients were subjected to phage therapy. Altogether 118 persons were included in the control group in which 74 lethal outcomes (62.7 %) were registered, while in the experimental group the mortality rate was almost one-tenth of the control group, with 5 cases out of 73 (6.8 %) [30].

In his book, d'Herelle mentioned the establishment of two industrial centers for the production of bacteriophages against cholera in 1931 in India [30, 37]. D'Herelle's commercial laboratory in Paris produced at least five phage preparations against various bacterial infections. The preparations were called *Bacté-coli-phage*, *Bacté-rhino-phage*, *Bacté-intesti-phage*, *Bacté-pyo-phage*, and *Bacté-staphy-phage*, and they were marketed by a company that later became the large French company L'Oréal [67–69].

---

[1] After the execution of George Eliava the book was impounded and access to it was limited to "for professional use only". N.Ch.

The Oswaldo Cruz Institute in Rio-de-Janeiro, Brazil started production of the anti-dysentery bacteriophages in 1924 to combat dysentery in Latin American countries [17, 30]. Within a year, the institute produced 10,000 vials of phages, which were sent into the hospitals around Brazil [29]. Therapeutic phages were also produced in the United States. In the 1940s, the Eli Lilly Company (Indianapolis, Ind.) produced seven phage products for human use, including preparations targeted against staphylococci, streptococci, *Escherichia coli*, and other bacterial pathogens [67]. These preparations consisted of phage-lysed, bacteriologically sterile broth cultures of the targeted bacteria (e.g., *Colo-lysate*, *Ento-lysate*, *Neiso-lysate*, and *Staphylo-lysate*) or the same preparations in a water-soluble jelly base (e.g., *Colo-jel*, *Ento-jel*, and *Staphylo-jel*). They were used to treat various infections, including abscesses, suppurating wounds, vaginitis, acute and chronic infections of the upper respiratory tract, and mastoid infections. However, the efficacy of phage preparations was controversial [18, 41]. This could be caused by the absence of viable phages, low phage titer, or narrow strain range of phage in these preparations, which was discovered in the case of some commercial anti-staphylococcal phage preparations [59]. As a result with the advent of antibiotics, commercial production of therapeutic phages ceased in most of the Western countries.

2.4 Mass Application of Phage-based Products in Medicine

From a review of historic literature, it is apparent that phage therapy trials were active in the 1930s and 1940s throughout Georgia, Russia, Ukraine, Belarus, and Azerbaijan, in the Soviet Union. Observations of cases associated with road accidents and septic infection carried out in hospitals in Moscow in [38–40] led to the development of methods and instructions for their intramuscular and even intravenous use, which was crucial in cases of generalized infections. The results of these observations were reported at conferences held in March, June, and December of 1940. These methods and instructions were approved by the Soviet Supreme Red Army Military-Sanitarian Office and were applied to the treatment of soldiers in the Red Army during the Finnish Campaign and the Second World War. The war and the need for therapeutic preparations inspired Soviet doctors to perform new trials with phages and to develop novel methods for treatment of wound infections. This period was one of the most fruitful in the development of phage therapy in the former Soviet Union [38–40, 70–72].

In 1920s and 1940s, the intestinal infections caused by *Salmonella* and *Shigella* species were a huge problem all over the world [2, 10, 14, 15, 18, 34, 40, 50, 52]. The mortality rate was very high [34]. Therefore, the efforts of Soviet scientists were focused on development of therapeutic phages for treatment of intestinal infections. The broadest clinical study on the therapeutic effect of dysenteric phages was reported by Sapir [62]. The author described altogether 1,064 cases of dysentery that were treated with bacteriophages in two different clinics in Moscow. The effect of phage therapy against intestinal diseases was confirmed by numerous authors [25, 27, 32, 47, 55, 76]. Phage preparations were generally considered to be particularly efficient for the treatment of intestinal infections. However, it was recognized that use of phage preparations is much more reasonable for prophylactic purposes. Therefore, the phage preparations have been used extensively in the Soviet Union for prophylaxis in the regions with high incidence of infections, and, also in the communities where rapid spread of infections might occur such as kindergartens, schools, and military accommodation, etc. [1, 3, 5–7, 19, 33, 35]. The preventive measures against bacterial epidemics performed by use of phages are known under the term "prophylactic phaging". Later on, mass prophylaxis of intestinal diseases was performed in the Red Army units by military doctors as well. For prevention of dysentery and typhoid epidemics, specific phages were also used with two tablets administered once every 5–7 days during the outbreak season [1, 3, 44].

2.5 Bacteriophages Versus Antibiotics

To evaluate effect of phage therapy over 5,000 volumes of scientific journals, selected articles, books and thesis of dissertations published between 1920s and 1980s have been screened. The gathered information was included into a monograph: "*A Literature Review of the Practical Application of Bacteriophage Research*" [11]. The main conclusions derived from the literature are: (a) Success outcome of phage therapy is up to 95 %; (b) Rapid improvement and cure is achieved within 3–5–7 days; (c) After phage therapy of wounds no scars are left (positive cosmetic effect); (d) No or rare relapsed cases are registered; (e) No side effects (allergy, yeast or fungal infections, kidney or liver failure, etc.) were registered; (f) Mortality rate among children suffering with septicemia was minimized; (g) Due to "prophylactic phaging" number of disease cases was reduces three to six times; (h) In the "phaged" group only mild disease cases were observed; (i) Phage therapy gives an effect similar to vaccination; (j) Phage therapy reduces the number of hospital days.

In general the difference between phages and antibiotics is based on a few characteristics. Phages are live bacterial viruses with specific action towards the bacterial targets and do not harm the useful bacteria that live in and on the body. However, antibiotics have generalized action that is often harmful for human and animal gut microbiota and may cause side effects. Since the phages are parasites they may reproduce themselves only in presence of appropriate hosts. The phages are active against drug-resistant bacteria. Since phage and bacteria are co-evolving prey and predator system, the phages may easily overcome phage-resistance of the host bacterial cell, if such occurs.

2.6 Phage Preparations

During the long history of using phages as therapeutic agents in the former Soviet Union and Eastern Europe, phages have been administered to humans (i) orally, in tablet or liquid formulations (10^5–10^7 PFU/dose), (ii) rectally, (iii) locally (skin, eye, ear, nasal mucosa, etc.), in tampons, rinses, and creams, (iv) as aerosols or intrapleural injections, and (v) intravenously, albeit to a lesser extent than the first four methods, and there have been virtually no reports of serious complications associated with their use.

Presently, the following phage preparations are in use in Georgia:

Pyo-bacteriophage: active against *Staphylococcus* sp., *Streptococcus* sp, *P. aeruginosa*, *Proteus mirabilis*, *Proteus vulgaris*, *E.coli* causing purulent infections. The preparation is recommended for treatment and prophylaxis purulent-inflammatory infections of ear, throat, nose, bronchus, lungs; surgical infections (abscesses, phlegmona, osteomyelitis, peritonitis); urogenital infections (urethritis, cystitis, pyelonephritis); gynecology infections (colpitis, endometritis, etc.); enteral infections (gastro-enterocolitis, cholecystitis, disbiosis), and purulent-septical infection of newborns.

Intesti-bacteriophage liquid: active against *Shigella* sp, *Salmonella* sp., *E.coli*, *Staphylococcus* sp., *Enterococcus* sp., *Proteus* sp., *Pseudomonas aeruginosa*. This preparation is recommended for treatment and prophylaxis of intestinal disorders, bacterial dysentery, shigellosis, salmonellosis, typhoid fever, paratyphoid salmonella, disbiosis, enterocolitis, colitis, and dyspepsia.

Enko-phage: active against the following species: *S. typhimurium*, *S. enteritidis*, *S. heidelberg*, *S. newport*, *S. cholerae suis*, *S. oranienburg S. dublin* and *S. anatum*; *Shigella flexneri* (serovars 1, 2, 3, 4) and *Shigella sonnei* (six different serovars), enteropathogenic *E.coli* (serovars: O11, O55, O26, O125, O113, O 128, O18, O44, O25, O20); *S. aureus*, *S. epidermidis* and *S. saprophyticus*.

SES-bacteriophage: active against Staphylococci (*S. aureus*, *S. epidermidis*, and *S. saprophyticus*); Streptococci (*S. pyogenes*, *S. sanguis*, *S. salivarius* and *S. agalactiae*) and different serotypes of enteropathogenic *E.coli* (O11, O55, O26, O125, O113, O 128, O18, O44, O25, O20).

Fersis–bacteriophage: active against Staphylococci (*S. aureus*, *S. epidermidis*, and *S. saprophyticus*) and Streptococci (*S. pyogenes*, *S. sanguis*, *S. salivarius* and *S. agalactiae*).

2.7 Intravenous Staphylococcal Bacteriophages (IVSP) for Treatment of Septic Infections

The development of apyrogenic Staphylococcus phage for intravenous use is one of the significant achievements of the Eliava Institute. This preparation is unique and does not have analogues in the world. Before its mass application, it was tested on animals, healthy human volunteers, and in clinical trials including up to 1000 patients (adults and children). For intravenous administration, the IVSP was used in a dose of 0.5–1 ml per 1 kg of weight as transfusions combined with blood replacing compounds (saline solution, etc.). Higher doses (2 ml per 1 kg of weight) were not usually applied except for in rare cases, for example, in the cases of osteomyelitis. The IVSP was also successfully applied in treatments and prophylaxis of post-traumatic or post-operational infections. For gynecological purposes, 10–20 ml of phage was administered locally into the infected sites. To avoid the difficulties related to the development of antibodies, the IVSP was administered once per day, every day, for 3–18 days depending on the severity of the disease and other factors. During the intravenous phage administration of the IVSP, the doctors did not observe any life threatening side effects [60].

Interestingly, in vitro screening performed between 2004 and 2009 in epidemiologically unrelated 352 *Staphylococcus* clinical strains using staphylococcal phage strain Sb-1 (known also as ISB), which is a key component of the intravenous commercial phage preparations (ISVP), demonstrated that 98–99 % phage-susceptibility of the clinical strains [46]. Methicillin Resistant Staphylococcus Aureus (MRSA) is one of the most important health concerns worldwide. Bacteriophage therapy is considered to be an alternative to antibiotics. Therefore, the efficacy IVSP was tested also against 424 human MRSA strains isolated in the UK. The results were promising, as 98 % of MRSA appeared to be susceptible to IVSP [23].

2.8 Current Status of Development of Phage Preparations and Their Applications

The Eliava Institute was established by d'Herelle and George Eliava as the world center of bacteriophage research and during its best times employed approximately 700–800 people, including researchers, technologists and support personnel. At that time, the institute produced phage preparations (often several tons a day) against a dozen bacterial pathogens, including intestinal bacteria, staphylococci, *Pseudomonas*, *Proteus*, and many enteric pathogens [21].

Production and use of phages for therapy and prophylaxis never stopped at the Eliava Institute of Bacteriophage, however, the scale is much smaller than before the break off of the Soviet Union. The therapeutic and prophylactic phages are presently produced by a small spin-off company Eliava BioPreparations, Ltd established by the Eliava IBMV. In the past the scale of the phage production covered the greatest part of the Soviet countries, including the Caucasian and Middle Asian republics. Today, production of the Eliava BioPreparations, Ltd fully satisfies the Georgian market. In 2010, the company started to export the phages outside Georgia into Azerbaijan.

These achievements would be impossible without intensive research held at the Eliava Institute of Bacteriophage, most of which are conducted in close collaboration with the international scientific community. This helps the Georgian scientists to develop further, introduce modern techniques into their research, and, what is the most important, to design proper experiments corresponding to western standards. On another hand, the collaborative projects help the western researchers to avoid 'discovery of wheel', which may easily occur without having a background knowledge in development of phage preparations and their application in practice. Over 50 collaborative international projects have been accomplished during the period of 1995–2014, which resulted in a number of significant publications [9–11, 24, 26, 43, 45, 51, 53, 57, 67, 74, 78].

Thanks to grant support, several new phage preparations have been developed during the recent decades, these are: PhageBioDerm, Uro-phage [51, 67], Oseteophage and Micolyse [12, 13].

The PhageBioDerm is a novel wound-healing preparation consisting of a biodegradable polymer impregnated with an antibiotic and lytic bacteriophages, and was recently licensed for sale in the Republic of Georgia. The "Uro-phage" has been successfully used in a clinical trial for treatment of chronic bacterial prostatitis (CBP) caused by Gram-positive (*Staphylococcus*, *Streptococcus* and *Enterococcus*) and Gram-negative (*E.coli*, *Pseudomonas* and *Proteus*) bacteria [36]. A composite for treatment of dental infections ("Oseteophage") was specifically formed by using a combination of hydraxyapatite, methyluracil, gelatin, and bacteriophages against *Staphylococcus*, *Streptococcus*, and *E.coli*. Clinical studies have approved that the preparation "Oseteophage" may be used to prevent dental archatropy and to stimulate bone regeneration [12]. Another novel bio-composite against oral infections caused by *Staphylococcus*, *Streptococcus*, and *Candida albicans*, and an antifungal extract of *Pupulus nigra* has been developed recently [13]. Gynecology is a promising area for application of bacteriophages therapy. A new combined preparation for treatment of bacterial and yeast diseases "Micolyse" was tested in a clinical experiment aimed at the treatment of pregnant women suffering from mixed bacterial and yeast infections.

Among other achievements in phage therapy, it is necessary to mention the treatment of cases of secondary infections associated with cystic fibrosis. Since 2008 phage preparations against secondary infections have been used in patients with cystic fibrosis. The study was performed in collaboration with the National Center of Cystic Fibrosis (Tbilisi, Georgia). Phages were applied to infants and adults via a nebulizer several times a day for 6–10 days. Simultaneously, patients were treated by conventional antibiotics, anti-mucus medications, and vitamins. Phage application caused a substantial decrease of bacterial counts in sputum samples and an improvement of the patient's general health condition. Due to bacteriophage therapy, long-term remissions of infections were achieved [43].

While the phage preparations are in everyday use in Georgia and in a number of ex-Soviet countries, there are many obstacles for the clinical application of bacteriophages in the western countries, such as the perception of viruses as 'enemies of life' [75]. This is due to the lack of a specific frame for phage therapy in the current Medicinal Product Regulation [57] and the absence of well-defined and safe bacteriophage preparations. To evaluate the safety and efficacy of bacteriophages in the treatment of burn wound infections in a controlled clinical trial, a highly purified and fully defined bacteriophage cocktail (BFC-1) was prepared, which is active against the *P.aeruginosa* and the *S.aureus* strains currently circulating in the Burn Centre of the Queen Astrid Military Hospital [53]. Based on successive selection rounds, three bacteriophages were retained from an initial pool of 82 *P. aeruginosa* and 8 *S. aureus* bacteriophages, specific for prevalent *P. aeruginosa* and *S. aureus* strains in the Burn Centre of the Queen Astrid Military Hospital in Brussels, Belgium. This cocktail, consisting of *P. aeruginosa* phages 14/1 (family *Myoviridae*) and PNM (family *Podoviridae*) and *S. aureus* phage ISP (family *Myoviridae*) was produced and purified of endotoxin. Quality control included stability (shelf life), determination of pyrogenicity, sterility, cytotoxicity, confirmation of the absence of temperate bacteriophages, and transmission electron microscopy-based confirmation of the presence of the expected virion morphologic particles, as well as of their specific interaction with the target bacteria. Bacteriophage genome and proteome analysis confirmed the lytic nature of the bacteriophages, the absence of toxin-coding genes showed that the selected phages 14/1, PNM and ISP are close relatives of respectively F8, φKMV and phage G1. The bacteriophage cocktail is currently being evaluated in a pilot clinical study was cleared by a leading Medical Ethical Committee. No adverse reactions were observed [53]. The detailed description of the quality controlled small-scale phage preparation was considered as a first step to promote the concept of phage treatment in Western medicine. In addition, it supported the creation of a discussion platform for regulatory framework for approval of phage therapy [57, 74].

Despite its long (Eastern European) history, phage therapy is not currently authorized for routine use on humans in the West. Today, it is only approved in some former Soviet republics like Russia and Georgia, where commercial phage preparations are sold in pharmacies [11]. In Poland, a recent member of the European Union, phage therapy is considered an 'Experimental Treatment' covered by the Physician Practice Act (Polish Law Gazette N° 28 of 1997) and the declaration of Helsinki, administered only when other therapeutic options do not exist [28, 57]. In France, therapeutic made-to-order phage preparations from the Institute Pasteur (Paris and Lyon) were used until the beginning of the 1990s. Today, a French practitioner, Alain Dublanchet, still uses commercial phage preparations (purchased in Russia and Georgia) to treat severe infections [57]. Despite the absence of a specific framework for phage therapy [28, 57], a pilot clinical trial in burn wounds was approved by an ethical committee in Belgium [45]. In the United States, a Food and Drug Administration (FDA)-approved phase I clinical trial was conducted. No safety concerns were found [45]. Recently, a British phage therapy company conducted a phase I/II a clinical trial in chronic otitis. This study was approved through the UK Medicines and Healthcare products Regulatory Agency (MHRA) and the Central Office for Research Ethics Committees (COREC) ethical review process [45].

Some bacterial pathogens are evolving a resistance to antibiotics, due to overuse and bacterial evolution. In addition, the antibiotic pipeline is running dry, with only a few new antibacterial drugs expected to make it to the market in the foreseeable future. Bacteria that are resistant to all available antibacterial drugs, so-called superbugs, are emerging worldwide. Evolutionary ecology might inform practical attempts to bring these pathogens under stronger human control [77].

Now, faced with the looming spectra of antibiotic resistance, Western researchers and governments are giving phages a serious look. In March, the US National Institute of Allergy and Infectious Diseases listed phage therapy as one of seven prongs in its plan to combat antibiotic resistance. Phagoburn: the first large, multi-center clinical trial of phage therapy for human infections, is funded by the European Commission, which is contributing €3.8 million (US$5.2 million) to the study.

Recently, the EU Parliamentary Assembly published the document entitled as 'Phage therapy, a public health issue' (Doc. 13480, 08 April 2014) (assembly.coe.int). In September 2014, a report to the President on combating antibiotic resistance and 'NATIONAL STRATEGY FOR COMBATING ANTIBIOTIC RESISTANT BACTERIA' were published in the USA as well (http://www.cdc.gov/media/dpk/2014/dpk-carb.html). These documents underline the growing threat of antibiotic resistance and call for better attention to bacteriophages therapy as a potential alternative to antibiotics.

Although governments are starting to pay attention to phage therapy, pharmaceutical companies remain reluctant to get on board. One of the reasons for this reluctancy is the intellectual property rights; even after almost 100 years history of phage therapy companies are still not allowed to claim a treatment as intellectual property and therefore recoup its costs. It is likely that a 2013 ruling by the US Supreme Court against the patenting of natural genes would also apply to phages isolated from nature. An engineered phage could, in theory, be patented. Recently the researchers started to use a DNA-editing system called CRISPR to kill only antibiotic-resistant bacteria. Its principle is based on the following: the phage injects the bacterium with DNA, which the microbe transcribes into RNA. If part of the bacterium's antibiotic-resistance gene matches that RNA sequence, an enzyme called Cas9 cuts up the cell's DNA, killing it. In initial trials, the researchers found that their phage could kill more than 99 % of the *E. coli* cells that contained specific antibiotic-resistance gene sequences, whereas it left susceptible cells alone. Giving the phage to wax worm larvae infected with resistant *E. coli* increased the worms' chance of survival. The researchers are now starting to test the system in mice (human trials are a long way off).

Phages have also emerged as promising tools for curing cancer. Thus, it was reported that phages play a potential protective role against oxidative stress [54]. Besides that, the engineered phages can be used as delivery vectors of anticancer agents [64]. It was shown that phages are inducing cytokines [58]. Phages may play a certain role for treatment of neurodegenerative diseases. Phage M13 has been used in treatment models for Alzheimer's and Parkinson's diseases because of its ability to bind to the typical b-amyloid and a-synuclein plaques in the brain, resulting in plaque disaggregation [42]. The human gut contains approximately 10^{15} bacteriophages (the 'phageome'), probably the richest concentration of biological entities on earth [16]. In Barr et al. [4] demonstrated that phages adhering to mucus provide a non-host-derived immunity to the human gut.

2.9 Phage-based Techniques for Identification of Bacteria

An increasing number of disease-causing bacteria are resistant to one or more anti-bacterial drugs utilized for therapy. Early and speedy detection of these pathogens is therefore very important. Traditional pathogen detection techniques, that include microbiological and biochemical assays, are long and labor-intensive, while antibody or DNA-based methods require substantial sample preparation and purification. Biosensors based on bacteriophages have demonstrated remarkable potential to surmount these restrictions and to offer rapid, efficient and sensitive detection technique for antibiotic-resistant bacteria.

Typical biosensor has three associated components: the sensor platform functionalized with a bio-probe to impart specificity of recognition; a transduction platform that generates a measurable signal in the event of analyte capture; the amplifier which amplifies and process the signal to give a quantitative estimate of analyte capture. Biosensors can be directly applied for the detection of pathogen in processed food matrices. Biosensors do not require the time-consuming sample during the pre-enrichment and secondary enrichment steps; therefore, this can accurately predict the amount and kind of food contamination much faster compared to conventional microbiological, immunological, and molecular biological methods [65].

Viruses have been used in a variety of chemical and biological sensors, and new applications continue to emerge. Bacteriophages, in particular, are one of the most successful viruses yet to have been applied towards biosensors, and can be grouped into three categories, regardless of their filamentous or icosahedral classification; one of these categories groups specific peptides or proteins discovered through phage display technology are genetically introduced onto the extraluminal surface of the bacteriophage for the detection of target analyte. In general, biosensors based on the use of nonlytic bacteriophages (i.e., M13) form the majority of this class. The next subgroup has specific bacterial cell markers that are detected upon bacterial cell rupture via the use of a lytic phage, which infects the host, replicates its genome, and induces host cell lysis. The last grouping holds bacteriophages that act a nano-scaffolding material for the transportation of functional molecules or their immobilization in a variety of sensing devices [48].

The requirements to sensor recognition elements are high stability, ease of immobilization on sensor platform, and recognition specificity towards host with minimum cross-reactivity from interfering pathogens. All of the three features are characteristic to bacteriophages. In different identification techniques phages have been applied either as whole phages, or phage-display peptides (PDPs), or as phage's receptor binding proteins (RBPs) [48].

Since it was first reported in 1985 in *Science*, one phage display technology has played a key role towards the development of antigen-specific peptides. Specifically, bacteriophages are used to choose select peptides with high specificity and affinity towards a desired antigen from a pool consisting of a large number of potential peptides. Their subsequent use for the recognition of specific target molecules and biomarkers has yielded a variety of economical, rapid, and efficient applications in fields such as vaccine development, enzyme inhibition, inflammation, and cancer research. In particular, the readily available information on the genomic makeup of various bacteriophages has enabled researchers to easily engineer foreign entities (i.e., proteins, peptides, and so on) on the extraluminal surface, which, in turn, has enabled their widespread use for phage display applications. While bacteriophages have continued to receive interest for these reasons, the last 10 years in particular have seen an emergence of novel bacteriophage-based biosensor applications in the form of nanotemplates or nanoprobes [48, 65].

Recently, the nanosized dimensions of bacteriophages, and ease with which genetic modifications can be made to their structure and function, have put them in the spotlight towards their use in a variety of biosensors. In particular, the expression of any protein or peptide on the extraluminal surface of bacteriophages is possible by genetically engineering the genome. In addition, the relatively short replication time of bacteriophages offers researchers the ability to generate mass quantities of any given bacteriophage-based biosensor. Coupled with the emergence of various biomarkers in the clinic as a means to determine pathophysiological states, the development of current and novel technologies for their detection and quantification is imperative [48].

Bacteriophages, conventionally used in phage display technology for the selection of peptides or antibody fragments, have emerged in the development of a variety of biomedical applications such as the fabrication of nanowire-based secondary battery cells, gene therapy constructs, sensors, and drug delivery vehicles.

Phage Receptor Binding Proteins (RBPs) are novel probes for pathogen detection. The unique host-specific recognition of the tailed phages comes from the RBPs located on the tail fibers and it is the binding of these proteins that trigger the translocation of the phage genetic material into the host. The phage's RBPs generally recognize unique proteins or carbohydrate (polysaccharide) sequences on the surface of the host bacterium. Genetically engineered RBPs offer several advantages over the antibody or intact phage-based technology for pathogen detection due to the following characteristics: their agglutination ability towards bacterial cells is found to be similar to the monoclonal antibodies against the bacterial lipopolysaccharides; the RBPs offer better stability against environmental factors such as pH and temperature and resistance against gastrointestinal proteases; and their binding affinity can be easily tailored to the requirement and multi-valency can be imparted, if desired [49].

The most commonly employed techniques to enhancement the effect of bacterial detection are: surface Plasmon resonance (SPR), Raman spectrometry, fluorescence/phosphorescence spectrometry, and Bio- or chem-iluminescence, orientated immobilization to different polymeric and/or metal surfaces, etc.[49, 65].

The translation of bacteriophage-based biosensors from in vitro to *in vivo* settings must be preceded by toxicity examinations to eliminate any safety concerns. Furthermore, the sensitivity of the general public to genetically engineered bacteriophages must be taken into account, and this is arguably one of the largest hurdles in the field. However, by coupling the aforementioned advantages of bacteriophages with the continued development of micro-and nano-based measurements and analytical devices, the scientists predict a concurrent increase in safety awareness and subsequent improvements. Finally, it is important to note that the use of bacteriophages is not limited to the aforementioned categories. Bacteriophages with varying functions that express desired peptides or proteins on their surface can be engineered through the simple manipulation of the bacteriophage genome, and ultimately be used for the rapid and economical mass production of bacteriophage-based biosensors. Here, an appropriate understanding of the bacteriophage that covers a wide area of topics, such as the structure and function of individual components and its life cycle, must accompany such novel applications. In summary, we envision that the combination of such bacteriophages with analytical nanofabricated devices will emerge in novel applications as a powerful nanoscale tool for the detection of a wide spectrum of targets, ranging from biological agents (i.e., proteins, bacteria, spores, viruses, toxins, and so on) to chemical agents or even explosives [49].

References

1. Agafonov BI, Khokhlov DT, Zolochevsky MA (1984) Epidemiology of typhoidparatyphoid infections and their prophylactics. Mil Med J 6:36–40
2. Alessandrini A, Doria R (1924) Il batteriofago nella terapia del tifo addominale. Policlinico Sez Prat 31:109
3. Anpilov LI, Prokudin AA (1984) Prophylactic effectiveness of the dry polyvalent dysentery bacteriophage in organized communities. Mil Med J 5:39–40
4. Barr JJ, Auro R, Furlan M, Whiteson KL, Erb ML, Pogliano J et al (2013) Bacteriophage adhering to mucus provide a non-host-derived immunity. Proc Natl Acad Sci U S A 110:10771–10776
5. Belikova MA (1941) Experience of phage prophilaxis of summer dysentery among the young children performed in the city Stalingrad. J Microbiol Epidemiol Immunol 5–6:142
6. Blankov BI (1941) Analysis of the results of phage prophylaxis of dysentery among the contacting people. Report # 1. J Microbiol Epidemiol Immunol 5–6:125–131
7. Blankov BI, Zherebtsov ID (1941) Experience on the multiple phaging of the contacting population in the fight against dysentery. Report # 2. J Microbiol Epidemiol Immunol 5–6:131–136
8. Bruynoghe R, Maisin J (1921) Essais de the‘rapeutique au moyen du bacteriophage. CR Soc Biol 85:1120–1121
9. Ceyssens P-J, Glonti T, Kropinski AM, Lavigne R, Chanishvili N, Kulakov L, Lashkhi N, Tediashvili M, Merabishvili M (2011) Phenotypic and genotypic variations within a single bacteriophage species. Virol J 8:134
10. Chanishvili N, Chanishvili T, Tediashvili M, Barrow PA (2001) Phages and their application against drug-resistant bacteria. J Chem Technol Biotechnol 76:689–699
11. Chanishvili N (2012) A literature review of the practical application of bacteriophages research. Nova Science Publishers, New York
12. Chkonia I, Meipariani A, Jgenti D, Alavidze Z, Goderdzishvili M, Dzidzishvili L, Kvatadze N, Mamamtavrishvili D, Menabde G (2009) Development of technology for creation "Osteophage" a composite stimulating bone regeneration. Proc Georgian Acad Sci Biol Ser B 7(1–2):63–69
13. Chkonia I, Alavidze Z, Goderdzishvili M, Meiphariani A, Dzizdzishvili M, Kvatadze N, Jgenti D, Makhatadze N, Rigvava S, Karumidze N, Kusradze I, Dvalidze T, Barbakadze S, Kuchukhidze J, Turabelidze D (2013) Natural phage biocomposite for mixed infections. Microbiol Biotechnol 3:10–12

14. Costa Cruz J (1924) Le traitement des dysenteries bacillaires par le bacteriophage. CR Soc Biol 91:845
15. Compton A (1929) Antidysentery bacteriophage in treatment of dysentery: a record of 66 cases treated, with inferences. Lancet 2:273
16. Dalmassso M, Hill C, Ross RP (2014) Exploiting gut bacteriophages for human health. Trends Microbiol 22(7):399–406
17. Dublanchet A, Bourne S (2007) The epic of phage therapy. Can J Infect Dis Med Microbiol 18:15–18
18. Eaton MD, Bayne-Jones S (1934) Bacteriophage therapy: review of the principles and results of the use of bacteriophage in the treatment of infections. JAMA 23:1769–1939
19. Florova NN, Cherkass FK (1965) Results of mass application of polyvalent dysenteric bacteriophage. J Microbiol Epidemiol Immunol 3:118–125
20. Gamaleya NF (1898) Bacterial lyzins: the enzymes destroying bacteria. Russ Arch Pathol Clin Med Bacteriol 6:607–613
21. Georgadze IA (1974) Fifty years of the Tbilisi Scientific-Research Institute of Vaccine and Sera of the Ministry of Health of the USSR. In: Selected articles of the jubilee dedicated to 50th anniversary of the Tbilisi Institute of Vaccine and Sera. TIVS, Tbilisi
22. Georgadze IA, Makashvili EG (1979) George Eliava. In: Abashidze I, Metreveli R (eds) The Georgian Soviet encyclopedia, vol 4. State Education Commission's Publishing House, Tbilisi, p 125
23. Giorkhelidze T, Koberidze T, Malkhazova I, Janelidze N, Tediashvili M, Chanishvili N, Hanlon G, Denyer S (2008) Screening results of the MRSA strains with Staphylococcus phages, phage, biology, ecology and therapy meeting, 12–15 June 2008, p 36
24. Glonti T, Chanishvili N, Taylor PW (2010) Bacteriophage-derived enzyme that depolymerizes the alginic acid capsule associated with cystic fibrosis isolates of *Pseudomonas aeruginosa*. J Appl Environ Microbiol 108:695–702
25. Gnutenko MP (1951) Treatment of *S. typhi* and *paratyphi* bacterial carriers with bacteriophages. J Microbiol Epidemiol Immunol 5:56–60
26. Goderdzishvili M, Meiphariani A, Chkonia I, Dzizdzishvili M, Kvatadze N, Jgenti D, Makhatadze N, Rigvava S, Karumidze N, Kusradze I, Dvalidze T, Gogbaidze G, Nikoladze B, Alavidze Z (2012) Development and initial testing of a phage cocktail to treat chronic bacterial prostatitis. Microbiol Biotechnol 3:15–18
27. Golubtsov GV (1940) Sero- and phage- therapies of dysentery among babies and infants. In: Selected articles of the 1st scientific conference of the Bashkir Medical University, Ufa, 23–25 Mar 1939, pp 13–16
28. Górski A, Miedzybrodzki R, Borysowski J, Weber-Dabrowska B, Łobocka M, Fortuna W, Letkiewicz S, Zimecki M, Filby G (2009) Bacteriophage therapy for the treatment of infections. Curr Opin Invest Drugs 10:766–774
29. Hausler T (2008) Viruses vs.superbugs. A solution to the antibiotics crises. Macmillan, London
30. d'Herelle F (1935) Bacteriophage and phenomenon of recovery. TSU Press, Tbilisi
31. d'Herelle F (1917) Sur un microbe invisible antagoniste des bacilles dysente'riques. CR Acad Sci (Paris) 165:373–375
32. Ionov ID, Erez SL, Goldenberg EY (1939) Specific therapy of dysentery. In: Variability of microorganisms and bacteriophage research (Proceedings of the scientific conference, 1936), Kiev, pp 337–346
33. Kagan MI, Kuznetsova EV, Teleshevskaya EA (1964) To the issue of epidemic effectiveness of the planned phaging in the day nurseries. J Microbiol Epidemiol Immunol 7:89–102
34. Karamov S (1938) Experience of phage therapy for treatment of typhoid fever. In: Selected articles of Azerbaijani Institute of Epidemiology and Microbiology, vol 6, Azerbaijan SSR, Baku, pp 101–105
35. Karpov SP (1946) The specific bacteriophage in relation to the issue of combating typhoid and paratyphoid diseases. J Microbiol Epidemiol Immunol 1–2:40–44
36. Karumidze N, Thomas JA, Kvatadze N, Goderdzishvili M, Hakala KW, Weintraub S, Alavidze Z, Hardies SC (2012) Characterization of lytic *Pseudomonas aeruginosa* bacteriophages via biological properties and genomic sequences. Appl Microbiol Biotechnol 94(6):1609–1617

37. Kazhal N, Iftimovich R (1968) From the history of fight against bacteria and viruses. Nauchnoe Izdatelstvo, Bucharest
38. Kokin GA (1941) Use of bacteriophages in surgery. Sov Med ("Sovietskaya Meditsina") 9:15–18
39. Kokin GA (1946) Phage therapy and phage prophylaxis of gas gangrene. In: Smirnov E, Grigolava S, Orbeli L (eds) Experience of the Soviet military medicine during the great patriotic war 1941–1945, vol 3. MedGiz Publishing House, Moscow, pp 56–63
40. Krestovnikova VA (1947) Phage treatment and phage prophylactics and their approval in the works of the Soviet researchers. J Microb Epidemiol Immunol 3:56–65
41. Krueger AP, Scribner EJ (1941) Bacteriophage therapy. II. The bacteriophage: its nature and its therapeutic use. JAMA 19:2160–2277
42. Ksendzovsky A, Stuart Walbridge BS, Saunders RC, Asthagiri AR, Heiss JD, Lonser RR (2012) Convection-enhanced delivery of M13 bacteriophage to the brain. J Neurosurg 117(2):197–203
43. Kutateladze M, Adamia R (2010) Bacteriophages as potential new therapeutics to replace or supplement antibiotics. Trends Biotechnol 28:591–595
44. Kurochka VK, Karniz AF, Khodyrev AP (1987) Experiences of implementation of preventive anti-epidemic measures in the center of intestinal infections with water transmission mechanism of morbidity. Mil Med J 7:36–37
45. Kutter E, De Vos D, Gvasalia G, Alavidze Z, Gogokhia L, Kuhl S, Abedon ST (2010) Phage therapy in clinical practice: treatment of human infections. Curr Pharm Biotechnol 11:69–86
46. Kvachadze L, Balarjishvili N, Meskhi T, Tevdoradze E, Skhirtladze N, Pataridze T, Adamia R, Topuria T, Kutter E, Rohde C, Kutateladze M (2011) Evaluation of lytic activity of staphylococcal bacteriophage Sb-1 against freshly isolated clinical pathogens. Microb Biotechnol 4(5):643–650
47. Lipkin YaO, Nikolskaya PV, (1940) Experience of phage therapy of dysentery. In: Selected articles of the Kuibishev Red Army Military-Medical Academy, vol 4. KRAMMA, Kuibishev, Russian SSR, pp 193–198
48. Lu TK, Bowers J, Koeris MS (2013) Advancing bacteriophage-based microbial diagnostics with synthetic biology. Trends Biotechnol 31(6):325–327
49. Lu TK, Koeris MS (2011) The next generation of bacteriophage therapy. Curr Opin Microbiol 14:524–531
50. Manolov DG, Sekunova VN, Somova EE (1948) Experience of therapy of typhoid fever by intravenous administration of the phage. J Microbiol Epidemiol Immunol 4:33
51. Markoishvili K, Tsitlanadze G, Katsarava R, Morris G, Sulakvelidze A (2002) A novel sustained-release matrix based on biodegradable poly (esteramide)s and impregnated with bacteriophages and an antibiotic shows promise in management of infected venous stasis ulcers and other poorly healing wounds. Int J Dermatol 41:453–458
52. Melnik MI, Nikhinson IM, Khastovich RI (1935) Phage prophylaxis of dysentery. In: Proceedings of the Mechnikov Institute in Kharkov, vol 1(1), MIKH, Kharkov, Ukrainian SSR89
53. Merabishvili M, Pirnay J-P, Verbeken G, Chanishvili N, Tediashvili M, Lashkhi N, Glonti T, Krylov V, Mast J, Van Parys L, Lavigne R, Volckaert G et al (2009) Quality-controlled small-scale production of a well-defined bacteriophage cocktail for use in human clinical trials. PLoS One 4:e4944
54. Miedzybrodzki R, Switala-Jelen K, Fortuna W, Weber-Dabrowska B, Przerwa A, Lusiak-Szelachowska M, Dabrowska K, Kurzepa A, Boratynski J, Syper D, Pozniak G, Lugowski C, Gorski A (2008) Bacteriophage preparation inhibition of reactive oxygen species generation by endotoxin-stimulated polymorphonuclear leukocytes. Virus Res 131(2):233–242
55. Mikaelyan VG (1949) Phage therapy of the bacterial carriers of typhoid bacteria. In: Proceedings of the Yerevan Medical Institute, vol 6. YMI, Yerevan, pp 54–59
56. Parfitt T (2005) Georgia: an unlikely stronghold for bacteriophage therapy. Lancet 365:2166–2167

57. Pirnay J-P, De Vos D, Verbeken G, Merabishvili M, Chanishvili N, Vaneechoutte M, Zizi M, Laire G, Lavigne R, Huys I, Van den Mooter G, Buckling A et al (2011) The phage therapy paradigm: prêt-à-porter or sur-mesure? Pharm Res 28:934–937
58. Przerwa A, Zimecki M, Switała-Jeleń K, Dabrowska K, Krawczyk E, Łuczak M, Weber-Dabrowska B, Syper D, Miedzybrodzki R, Górski A (2006) Effects of bacteriophages on free radical production and phagocytic functions. Med Microbiol Immunol 195:143–150
59. Rakieten ML (1932) Studies with Staphylococcus bacteriophage. I. The preparation of polyvalent Staphylococcus bacteriophage. Yale J Biol Med 4:807–818
60. Ministry of Health, Ministry of Medical Production of the USSR, Chief Agency of Microbiological Production (1985) Report of the clinical trials.
61. Rice TB (1930) Use of bacteriophage filtrates in treatment of suppurative conditions: report of 300 cases. Am J Med Sci 179:345–360
62. Sapir IB (1939) Observations and recommendations related to phage therapy of dysentery. In: Proceedings of the Moscow Institute of Infectious Diseases after I.I. Mechnikov, pp 135–151
63. Schless RA (1932) *Staphylococcus aureus* meningitis: treatment with specific bacteriophage. Am J Dis Child 44:813–822
64. Shoae-Hassani A, Sharif S, Verdi J. (2011) The neurosteroid dehydroepiandrosterone could improve somatic cell reprogramming. Cell Biol Int. 35(10):1037–41. doi: 10.1042/CBI20100927
65. Singh A, Poshtiban S, Evoy S (2013) Recent advances in bacteriophage based biosensors for food-borne pathogen detection. Sensors 13:1763–1786
66. Stout BF (1933) Bacteriophage therapy. Tex State J Med 29:205–209
67. Sulakvelidze A, Alavidze Z, Morris JG Jr (2001) Bacteriophage therapy. Antimicrob Agents Chemother 45:649–659
68. Summers WC (1999) Felix d'Herelle and the origins of molecular biology. Yale University Press, New Haven
69. Summers WC (2001) Bacteriophage therapy. Annu Rev Microbiol 55:437–451
70. Tsulukidze AP (1938) Application of phages in urology. Urology ("Urologia") 15:10–13
71. Tsulukidze AP (1940) Phage treatment in surgery. Surgery ("Khirurgia") 12:132–133
72. Tsulukidze AP (1941) Experience of use of bacteriophages in the conditions of war traumatism. Gruzmedgiz, Tbilisi
73. Twort F (1915) An investigation on the nature of ultramicroscopic viruses. Lancet 11:1241
74. Verbeken G, Huys I, Pirnay J-P, Jennes S, Chanishvili N, Scheres J, Górski A, De Vos D, Ceulemans C (2014) Taking bacteriophage therapy seriously: a moral argument. BioMed Res Int 2014:1–8
75. Villareal LP (2005) Overall issues of viral and host evolution. In: Villareal IP (ed) Virus and evolution of life. ASM Press, Washington, DC, pp 1–28
76. Vlasov KF, Artemenko EA (1946) Treatment of chronic dysentery. Sov Med ("Sovetskaya Medicina") 10:22–28
77. Williams PD (2010) Darwinian interventions: taming pathogens through evolutionary ecology. Trends Parasitol 26:83–92
78. Yilmaz C, Colak M, Yilmaz BC, Ersoz G, Kutateladze M, Gozlugol M (2013) Bacteriophage therapy in implant-related infections: an experimental study. J Bone Joint Surg Am 95(2):117–125

Chapter 3
Xanthomonas vesicatoria Specific Virus and Its Potential to Prevent Tomato Bacterial Spot Disease

Tinatin Sadunishvili, Edisher Kvesitadze, and Giorgi Kvesitadze

Abstract The effect of prolonged and over usage of chemicals in crops production has resulted in human health hazards and pollution of environment and ground water. Identification of new sources for biological control of plant diseases is important for sustainable agriculture, ensuring food security, improving human health and rehabilitating the environment. The use of bacterial viruses or bacteriophages for bacterial diseases control is a fast expanding area of plant protection. Study of phages diversity, specificity, stability and efficacy are important for their application as biological means against the pathogens. The paper summarizes data on properties of bacteriophages specific to *Xanthomonas vesicatoria* strains spread in Georgia and efficacy to prevent tomato bacterial spot in laboratory conditions under artificial infection.

3.1 Introduction

Food crops are susceptible to diseases worldwide caused by bacteria. As a result of bacterial diseases the loss of different kinds of vegetables in the field as well as post-harvest is high and varies depending on disease and geographic location. When the conditions are beneficial for phytopathogen multiplication, the loss may be very high, significantly affecting country's economics [11].

T. Sadunishvili
Durmishidze Institute of Biochemistry and Biotechnology, Agricultural University of Georgia, 240 David Agmashenebeli Alley, 0159 Tbilisi, Georgia

E. Kvesitadze
Georgian Technical University, 77 Kostava St., 0168 Tbilisi, Georgia

G. Kvesitadze (✉)
Durmishidze Institute of Biochemistry and Biotechnology, Agricultural University of Georgia, 240 David Agmashenebeli Alley, 0159 Tbilisi, Georgia

Georgian National Academy of Sciences, 52 Rustaveli Ave., 0108 Tbilisi, Georgia
e-mail: kvesitadze@hotmail.com

T.A. Camesano (ed.), *Nanotechnology to Aid Chemical and Biological Defense*, NATO Science for Peace and Security Series A: Chemistry and Biology,
DOI 10.1007/978-94-017-7218-1_3

Crop diseases caused by bacteria need to be controlled to maintain the quality and abundance of food. Different approaches may be and are used to prevent, mitigate or control plant diseases [36].

Application of pesticides against plant pathogens and overuse of chemical fertilizers together with good agronomic practices have contributed significantly to the spectacular improvements in crop productivity and quality over the past century. However, excessive use and misuse of agrochemicals has led to environmental pollution and consequently to considerable changes in society's attitudes towards the use of pesticides in agriculture. Development of alternative methods for controlling of pests and diseases is of great importance.

Among various diseases caused by fungi, viruses, nematodes and bacteria, bacterial spot caused by the Gram-negative phytopathogenic bacterium *Xanthomonas campestris* pv. *vesicatoria* is the most devastating in Georgia and worldwide. The disease can cause severe losses in tomato crop yield and fruit quality, which is generally unmarketable and negatively affects the economics of the country. It is highly destructive both in greenhouses and field grown tomatoes causing 10–50 % yield loss in many tomato production areas. For instance, in Georgia, Florida and the Caribbean this disease may cause decrease in tomato productivity up to 50 %, while in some Canadian fields loss of productivity can reach 60 % [8, 31].

Xanthomonas vesicatoria infects tomato seeds, all aboveground organs – stems, foliage, flowers and fruits. This bacterium cause injures of tomato leaves both in young plants and in fully-grown ones, where dark-colored lesions trend to concentrate at the edges of blades are formed. The leaves resemble sunburns (heliosis) ranging as from light-colored spots to complete defoliation. A secondary rot may develop as well [29].

X. vesicatoria is able to survive on tomato volunteers and plant debris from infected tissue. Seeds also serve as a medium for survival and dissemination of this bacterium. Disease development is favored by temperatures of 24–30 °C and by high humidity and rain. The dissemination of this bacterium occurs by wind driven rain or irrigation droplets as well as some cultural practices. *X. vesicatoria* enters through natural openings (stomata) and wounds created by mechanical means, wound-driven sand or insect punctures [22].

Bacteria of *X. vesicatoria* maintain their vital activity (pathogenicity) during 2 years after appearing on seeds; they are transmitted from infected seeds to robust ones, sowing of which cause lack of tomato plants and yield on planted area resulting in significant economic damage.

Currently, management of tomato bacterial spot is mainly achieved by integration of cultural practice, chemical control and resistance breeding. Cultural practice, such as application of disease- and pathogen-free transplants, repeated digging, burning of plant wastes, 3 years or longer crop rotation are poorly low effective. Copper salts, especially the water insoluble, so called "fixed coppers," are the most effective mean for treatment of aboveground parts of plants. However, feasibility and reliance on copper containing pesticides is questioned because of the emergence of tolerant strains, phytotoxicity, and soil contamination, especially when applied at high concentrations to overcome the resistance and unfavorable weather conditions [26].

With the aim of environmental protection and production of ecologically pure food, creation and application of biological pesticides against plant bacterial diseases is extremely important.

Antibiotics produced by microorganisms have been shown to be particularly effective suppressing plant pathogens and the diseases they cause. However, the use of antibiotics had been mostly discontinued due to the emergence of antibiotic-resistant bacteria [41]. The antibiotic resistance is not the only challenge for the use of these chemicals against phytopathogens. Another serious concern with antibiotics is spread of resistance genes to other bacteria, including other pathogenic or nonpathogenic bacteria present in the environment [32].

Many microorganisms produce and release lytic enzymes that can hydrolyse a wide variety of polymeric compounds, such as chitin, proteins, cellulose, hemicellulose, DNA. Microorganisms that show a preference for colonizing and lysing plant pathogens (by means of antibiotics or lytic enzymes they produce) might be classified as biocontrol agents. At the same time antibiotics are toxic to a verity of living organisms, including human beings. There are no data on specific antibioltics and/or lytic enzymes that expressed specific effect to *Xanthomonas* species.

Phytoncids and phytoalexins that are biologically active substance of plant origin kill or inhibit the growth and development of bacteria, microscopic fungi, and protozoa. Phytoncides play an important role in plant immunity and in the interrelationships of organisms in biogeocoenoses. The antimicrobial potency and range of phytoncides vary greatly among different species of plants and study on their potential against phytopathogens of vegetables could be of interest [20].

Chemicals that activate plant defense responses have been studied to apply as alternative chemical control approaches. Harpin (Messenger, Eden Bioscience, WA) and acibenzolar-S-methyl (ASM), known as Actigard, Syngenta NC are such chemicals that induce systemic acquired resistance of plants to bacterial infection. Application of the ASM have shown activity against bacterial spot in tomato [30].

One of the alternative ways of bacterial diseases prevention could be the use of genetically modified plants. However, there is consumer reluctance to use them as food in many countries.

Biological control of tomato bacterial spot has been achieved by using the nonpathogenic T3 strains to antagonize the pathogenic T1s in Florida [19, 23] and with other antagonistic bacteria such as *Pseudomonas putida* B56 and *Pseudomonas syringae* Cit7 in case of tomato bacterial speck [43].

Another approach for biological control of plant bacterial diseases is the use of bacterial viruses, called bacteriophages or phages. Bacterial viruses are the largest biological group in nature [38]. Over 5500 different bacteriophages have been discovered to date, each of which being able to infect one or several types of bacteria [1]. Phages have been proposed as plant-pathogen control agents in phage therapy [12]. Lately interest in phages has been renewed due concerns over environmental contamination [5]. Due to its high specificity it is assumed that phages are non-toxic to the eukaryotic cells [13] and thus no adverse effects to humans, wildlife or the environment are expected from them. The main advantages of phages are that they are specific for the disease-causing bacterium; they are self-replicating and self-limiting [27]. Phage treatment, integrated with other practices (plant activators, recent formulated phages in combination with other biological control agents and systemic acquired resistance inducers) is currently used in greenhouses and fields [24].

Bacteriophages were used as biocontrol agents as early as 1926 [34], and were effective in controlling Stewart's disease in corn (*Zea mays* L) in 1935 [42]. However, use of phages was abandoned due to the emergence of bacterial mutants resistant to the single-phage types used.

A method to reduce the likelihood of phage-resistant bacterial mutants emerging in a cropping system was developed by Jackson [21]. He used a mixture of different phages, including host range (h-) mutant bacteriophages. H-mutant phages, capable of attacking an extended range of bacterial isolates within a species, are spontaneously derived from their wild type parent phages, and are so named because they lyse not only parent wild-type bacteria, but also phage-resistant mutants originating from parent bacteria [10].

Paper presents results on study of physical-chemical and biological properties of spread in Georgia *Xanthomonas vesicatoria* specific bacteriophage and its potential for tomato bacterial spot prevention.

3.2 Bacteriophage Isolates

Dozens of *X.vesicatoria* strains were isolated from diseased tomato plants collected in different tomato production sites in Georgia. Their virulence, cultural-morphological and physiological properties have been studied. *X.vesicatoria* strains on which the phages displayed high activities were applied in experiments and none of them were lysogenic [14]. As a result of study of the *X.vesicatoria* strains on bacteriocin production one strain with such ability was excluded from further experiments.

Isolation of bacteriophages have been carried out from sewage and diseased tomato plants [3] collected in different tomato production regions in Georgia.

As a result six filtrates have been isolated and their activities to standard and freshly isolated strains were checked [14].

Lytic activities of the six filtrates were enhanced by multiple passaging. Their host range studies show that different phages lysed different amount of tested strains ranging from 48 % to 78 % (Table 3.1).

3.3 Phage Pure Lines

With this aim phages pure lines have been isolated from the mixture of six filtrates with similar titer by double layer agar method. The mixture of phages lysates displayed round-shaped light plaques of different sizes: large plaques (d = 4–5 mm), midsize plaques (d = 2–3 mm) and small ones (d = 1 mm). After multiple passages

Table 3.1 Host range of *X.vesicatoria* bacteriophages

		Amount of lyzed strains	
Phage #1	*X.vesicatoria* strains tested	Number	%
1	73	38	52
2	73	49	67
3	73	35	48
4	73	42	58
5	73	51	70
6	73	57	78

Table 3.2 Lytic activities of phage lines and their mixture to *X.vesicatoria* strains

		Phage pure line, #		
Strain	Polyclonar phage	1	2	3
103	4×10^5	3×10^2	–	2×10^2
1166	9×10^{10}	4×10^8	5×10^5	7×10^6
1225	8×10^{11}	4×10^{10}	3×10^{11}	7×10^{11}
1204	6×10^6	3×10^4	–	8×10^3
1171	5×10^{10}	8×10^9	7×10^9	3×10^{10}
1141	4×10^8	4×10^4	8×10^6	3×10^5
1167	5×10^6	3×10^4	9×10^5	6×10^4
1242	2×10^7	8×10^4	5×10^6	9×10^4
1258	2×10^7	9×10^6	8×10^6	1×10^7
1272	3×10^{10}	6×10^{10}	8×10^8	2×10^9
7	4×10^6	–	2×10^6	–
1235	3×10^8	8×10^6	4×10^6	2×10^7

three phage pure lines from all the three sized plaques were obtained. However, later they all obtained from morphologically homogenous plaques, regarding the host-cell displayed similar three size plaques, supposedly indicating on polymorphic nature of the phages. It should be mentioned that the sensitivity of *X.vesicatoria* strains to phage pure lines were different [15].

Similarly, the three phage pure lines displayed different lytic activities towards different host cells in a range 10^2–10^{11} PFU (Table 3.2) thus indicating on different yield of phage particles from bacterial cells. At the same time all three phages and their polyclonar mixture were characterized by high titers on most of the strains and thus they can be considered as phages with high yield on their host cells.

3.4 Bacteriophage and Host Bacteria Interaction

Biological properties, in particular productive life cycle of bacteriophages, i.e. development inside the bacterial host have been studied. Adsorption rate, latent period and phage yield are the values that characterize bacteriophage and bacteria host interactions.

Table 3.3 *X. vesicatoria* specific phages intracellular propagation

Phage	*X. vesicatoria* tested strains total number	Adsorbtion at 28 °C	Cycle of development inside strain at 28 °C		
		Time, min	%	$K10^{-9}$ min^{-1}	Latent period, min
Phage 1	18	21–24	86	2.6–2.3	36–45
Phage 2	18	21–24	83	2.2–1.9	36–45
Phage 3	18	18–21	91	4.7–4.0	36–42

Initial titer of phages: phage 1–4×10^8 PFU/ml, phage 2–3×10^8 PFU/ml,phage 3–7×10^8 PFU/ml. Initial titer of bacterial culture – 3×10^8 CFU/ml

Adsorption rate of phage on bacterial cells was determined by means of chloroform method, according to the amount of unabsorbed phage particles; the constants of adsorption rate were calculated according to the equation [17]

$$K = 2.3/\,(B \times t) \times \lg\,(Po/P)\,.$$

It was established that the adsorption period of three phage lines on tested bacterial cells varied from 18 to 24 min, adsorption constants in a range $2.3 - 4.7 \times 10^{-9}$ min^{-1} (Table 3.3).

Determination of the phages latent period was conducted by counting of time of appearance of large amount of phage particles from the time when amount of unabsorbed phage became constant by applying of double-layer agar method. The latent periods for the phages varied between 36 and 45 min (Table 3.3).

Thus, adsorption time and latent period were short and nearly identical for the phages of the three lines. These properties together with high yield of the lytic phages on their host bacterial cells are those that are required for their application in phage therapy.

3.5 Frequency of Development of Phage Resistant Strains

Resistance frequency of *Xanthomonas vesicatoria* bacterial strains to specific bacteriophages has been studied. It is important to note that the rate of development of resistant bacteria mutants to their viruses highly determines efficacy of the later as to control the pathogen and disease.

The frequency of mutation of phage resistant *Xanthomonas vesicatoria* bacteria strains were studied in different variations were individual phages and their mixture were applied in different titers, as well as the ratio of three phages in their mixtures were changed.

Determination of resistance frequency was conducted with phages three pure lines and their mixture on *X. vesicatoria* bacteria strains (strains # 7, #1225 and #1241 and the standard strains #1166 and #1258).

Table 3.4 Rate of development of resistant mutants of *Xanthomonas vesicatoria* bacterial strains to bacteriophages

		Phage titre 10^6 PFU/ml		
		Bacteria titre 10^9 CFU/ml		
Strain No	Phages mixture 1:1:1	1 (Large plaques)	2 (Medium plaques)	3 (Small plaques)
1241	10^4	10^2	10^3	10^4
1225	10^4	10^3	10^3	10^4
1258	10^4	10^3	10^4	10^4
1166	10^4	10^2	10^4	10^4
7	10^4	10^3	10^3	10^4

Table 3.5 Rate of development of resistant mutants of *Xanthomonas vesicatoria* bacterial strains to bacteriophages

		Phage titre 10^8p.f.u./ml		
		Bacteria titre 10^8c.f.u./ml		
Strain No	Phages mixture 1:1:1	1 (Large plaques)	2 (Medium plaques)	3 (Small plaques)
1241	10^7	10^5	10^5	10^7
1225	10^7	10^6	10^7	10^7
1258	10^7	10^6	10^6	10^7
1166	10^7	10^5	10^5	10^6
7	10^7	10^6	10^7	10^7

It was shown that at low titer 10^6 PFU/ml of phages and that of bacteria 10^9 CFU/ml, resistant bacterial strains development frequency was high, in a range 10^2–10^4 with comparatively lower indices 10^4 for Phage 3. It should be mentioned that the development of resistant mutants was low to phage mixture at ratio: 1:1:1 (Table 3.4).

When phages were applied in higher titer of 10^8 PFU/ml, similar to bacterial concentration 10^8 CFU/ml resistance frequency was rather low 10^5–10^7 (Table 3.5).

It should be noted that the lowest frequency of resistant mutants development was observed to phages mixture as compared to individual phages (Tables 3.4 and 3.5). In experiments with phage mixtures applied at different ratios (2:1:1, 1:2:1, 1:1:2), under conditions of phage and bacteria titers as in previous case, in all variations phages mixtures were characterized by the low frequency of resistance development, corresponding to 10^6–10^7. The lowest mutation rate 10^7 was observed for case where phages were applied at the ratio 1:1:2. Not surprisingly, as the phage 3 was characterized by lower frequency of resistance development. Similar indices for the mixture of the three phages at ratio 1:1:1 is evidence of the prevalence of application of phage mixture to lyse host bacterial cells rather than individual phages.

Development of resistance frequency of bacteria is an important precondition for creating of polyvalent bacteriophage preparation since the low index of phage resistance is the evidence of phage activity.

Thus the three phages and their mixtures are characterized by the low frequency of development of resistant bacteria cells and could be successfully applied against these bacteria. The optimum ratio of the phages for polyvalent preparation was established 1:1:1.

3.6 pH and Temperature Stability of the Bacteriophages

In order to study the influence of pH on bacteriophage, the later (10^8 PFU/ml) was incubated in the 0.1 M Tris-acetate buffer (pH 2.5–9.0) and glycine-NaOH buffer (pH 10.0–12.5) for 1 h. Phage activity was determined by applying of double-layer agar method [3, 27]. It was shown that optimum pH for the pages activities was in range – pH 6.8–7.2. Number of phage particles decreased significantly at pH below 5.5 and above 8.0.

To study the temperature stability, all three phage clones and polyvalent preparation were incubated at different temperatures from 55 °C to 82 °C during 120 h. All three phages and their mixture retained their titer after 120 min incubation at 65 °C. Thirty minutes incubation at 70 °C temperature resulted in twofold decrease of the phages titer, while 2 h exposure lad to complete inactivation. Higher – 82 °C temperature inactivated the phages mixture as well as phage separate clones in 10 min. Thus, the phages could be characterized as highly thermostabile phages, that is important for their possible application as biopesticides.

3.7 Phage Efficacy to Control Tomato Bacterial Spot

As was mentioned above *X.vesicatoria* can infect all aboveground parts of tomato including the seeds. They retain their pathogenic properties during 2 years and may be spread to healthy seeds. Sowing of such seeds results in significant decrease of tomato seedlings and economic loss.

The efficacy of the polyvalent bacteriophage preparation to prevent tomato seedlings from bacterial spot was studied. The selected healthy seeds (washed and dryed) were soaked in washed out 18–24-h bacterial culture (5×10^8 CFU/ml) grown on 2 % potato agar. After 1 h incubation seeds were sprayed with phage 10^7 PFU/ml and sowed in pots as other control seeds (intact and bacteria infected ones). Optimum temperature (24–28 °C), good aeration and high humidity was maintained in the greenhouse. After 5 weeks seedlings were analyzed (Fig. 3.1).

The treatment of infected seeds with polyvalent phage preparation resulted in twofold increase of raised seedlings with well developed shoots and roots as compared to bacterial infected plants (Table 3.6).

Efficacy of the developed polyvalent bacteriophage preparation against tomato bacterial spot was studied on artificially infected by *X. vsicatoria* seedlings and green fruits. One time spraying of polyvalent phages mixture prevented tomato

Fig. 3.1 Effect of phage treatment on development of tomato seedlings from artificially infected seeds. *I*- Intact seedlings; *IIa*-Seedlings raised from infected by *X. vesicatoria* strain 7 seeds; *IIb*- Seedlings raised from infected by *X. vesicatoria* strain 1258; *IIIa*- Seedlings raised from infected by *X.vesicatoria* strain7 and treated with phage; *IIIb*- Seedlings raised from infected by *X. vesicatoria* strain 1258 and treated with phage

Table 3.6 Rais of tomato seedlings from artificially infected by *X. vesicatoria* and phage treated seeds

Variant of experiment	Number of raised seedlings out of 50 seeds	Survival, %	Length of shoots, cm	Length of roots, cm
Intact seeds	42	84.3	19	3.5
Seeds infected by *X. vesicatoria* strain 7	18	35.7	11	1.2
Seeds infected by *X. vesicatoria* strain 1258	16	31.4	9	1.0
Seeds infected by *X. vesicatoria* strain 7 and treated with phage	33	65.7	16	2.5
Seeds infected by *X. vesicatoria* strain 1258 and treated with phage	31	61.4	15	2.5

plants from bacteria spot when it was applied at the moment or after the 24 h of inoculation of foliages with bacterial culture; spraying of the phage preparation a week later after inoculation, resulted in the decrease of disease severity and plant recovery [16]. Similar results were obtained in case of artificial infection of tomato green fruits [37].

Polyvalent bacteriophage preparation, specific to *X. vesicatoria* Georgian population, causing tomato bacterial spot could be successfully used for biological control of the disease. Treatment of tomato plants and seeds either with preventive or cure purposes exhibited high efficacy in greenhouse conditions.

3.8 Prospects and Challenges

Bacterial viruses, which have a lytic cycle, as opposed to lysogeny, are the ideal candidates for biopesticides. Lytic phages adsorb on host pathogens and cause their lysis after immediate replication of the virion.

Great genetic diversity of bacterial strains causing a bacterial disease of a certain plant is a major factor for identifying suitable phages for biocontrol. Application of specific to target bacteria lytic phages in a phage coctail results in significantly improved control efficacy [39]. The control effect can be increased even further if applying a mixture of host-range mutant phages [9]. Strategy for minimization of the occurrence of phage-resistant mutants was developed by Jackson [21]. By preparing mixtures of wild-type phages and including host range mutant phages (h-mutants), bacterial strains resistant to the parent phage are lysed. In field trials, tomato bacterial spot control with the mixture of four phages including wild-type and h-mutant phages when applied twice weekly to plants provided significantly better disease control and produced greater yield of extra large tomato fruits than the standard copper-mancozeb treatment [9].

Phage concentrations is of significance importance for disease control. Balogh et al. [4] observed that phage mixtures applied at 10^6 or 10^8 PFU/ml concentration provided similar levels of control of bacterial spot to tomatoes inoculated with10^8 CFU/ml of *Xantomonas perforans*, but at 10^4 PFU/ml was ineffective.

Several environmental factors were analyzed as potential inhibitors of the disease control by phage in phylosphere and rhyzosphere [6, 12]. Persistence on leaf surfaces is a major limiting factor in using phage therapy for disease control in the phyllosphere. Both laboratory and field studies have demonstrated that several factors as high temperatures, high and low pH and sunlight, rain are critical and determinant of the short-lived persistence of phages on plant leaf surfaces. These factors in addition to UV-A and UV-B spectra of sunlight the most destructive factors to viruses in common, are the major limiting factors for phage therapy in the phyllosphere.

Several strategies have been evaluated for increasing phage persistence [2, 7], including the use of protective formulations, proper frequency and timing of application scheduling for sunlight avoidance, and co-application of bacterial hosts for in vivo phage propagation. The protective formulations have been developed [35] containing milk, sugar and flour [2] in addition to selected phage preparation, that succeeded in elimination of the reduction of control efficacy caused by unfavorable environmental factors [20].

In addition to environmental factors, the abundance of host bacteria causes fluctuations in phage numbers [18]. On leaf surfaces, where high host populations persist, phages persist at significantly higher levels than on surfaces without the host.

For this purpose, avirulent strain to which phage is active could be used, as described for *Ralstonia solanacearum* to reduce tobacco bacterial wilt incidence [40].

Phages can be used effectively as part of integrated disease management strategies. Obradovich et al., formulated phages combined with other biological control agents and plant resistance inducers [35]. Phage-based integrated management of tomato bacterial spot is now officially recommended to growers of tomato in Florida [33] and relevant bacteriophage mixtures (Agriphage from OmniLytics Inc., Salt Lake City, UT EPA Registration #67986-1) are commercially available. Agriphage represents a mixture of active bacteriophages specific to *Xanthomonas campristris* pv. *vesicatoria* and *Pseudomonas syringe* and was registered first times in 2005 (www.omnilytics.com).

Selection of a suitable phage is a critical factor in ensuring success of phage therapy in agriculture. Great genetic diversity of bacterial strains causing a bacterial disease of a certain plant is a major factor for identifying suitable phages for biocontrol. For this, in vitro assays only are not adequate predictors of biological control ability (See for review [25]). For example, plaque size, antibacterial activity or phage multiplication rate for eight *Xanthomonas perforans* phages, was not correlated with disease control efficacy. Therefore, actual plant bioassays are unavoidable in order to gauge biocontrol activity. Bacterial strains may vary within a species in their sensitivity to bacteriophage. So, phage selection for field use requires careful monitoring of bacterial strains in the field for their natural resistance to deployed phages. To reduce the evolution of resistant bacteria, further developments are needed. They may include application of phages in combination with other antimicrobials, such as antibiotics, cycling through different phage mixtures, and engineering phages to directly target phage-resistance mechanisms [28].

Acknowledgments This work was supported by ISTC G-1129 and GNSF-STCU 5001 grants.

References

1. Ackermann HW (2007) 5500 phages examined in the electron microscope. Arch Virol 152:227–243
2. Balogh B, Jones JB, Momol MT, Olson SM, Obradovic A, King P et al (2003) Improved efficacy of newly formulated bacteriophages for management of bacterial spot on tomato. Plant Dis 87:949–954. doi:10.1094/PDIS.2003.87.8.949
3. Adams MH (1959) Bacteriophages. Interscience Publishers, New York
4. Balogh B, Jones JB, Momol MT, Olson SM (2005) Persistence of bacteriophages as biocontrol agents in the tomato canopy. Proc Int Symp Tomato Dis, 1st, Orlando, FL.ISHS Acta Hortic 695:299 101–101
5. Balogh B (2006) Characterization and use of bacteriophages associated with citrus bacterial pathogens for disease control. PhD thesis University of Florida, Gainesville, 112p
6. Basit HA, Angle JS, Salem S, Gewaily EM (1992) Phage coating of soybean seeds reduces nodulation by indigenous soil bradyrhizobia. Can J Microbiol 38:1264–1269
7. Bergamin FA, Kimati H (1981) Estudos sobre um bacteriofago isolado de *Xanthomonas campestris*. II. Seu emprego no controle de *X. campestris* e *X. vesicatoria*. Summa Phytopathol 7:35–43

8. Bouzar H, Jones JB, Stall RE, Louws FJ, Schneider M, Rademaker JLW et al (1999) Multiphasic analysis of *Xanthomonads* causing bacterial spot disease on tomato and pepper in the Caribbean and central America: evidence for common lineages within and between countries. Phytopathology 89:328–335
9. Flaherty JE, Jones JB, Harbaugh BK, Somodi GC, Jackson LE (2000) Control of bacterial spot on tomato in the greenhouse and field with H-mutant bactetriophages. Hortic Sci 35:882–884
10. Flaherty JE, Jones JB, Harbaugh BK, Somodi GC, Jackson LE (2001) H-mutant bacteriophages as a potential biocontrol of bacterial blight of geranium. HortScience 36:98–100
11. Frampton RA, Pitman AR, Fineran PC (2012) Advances in bacteriophage mediated control of plant pathogens. Int J Microbiol 13:326452, 11 pages
12. Gill JJ, Abedon TS (2003) Bacteriophage ecology and plants APSnet feature. http://www.apsnet.org/online/feature/phages/abedon.pdf
13. Greer GG (2005) Bacteriophage control of foodborne bacteria. Food Prot 68:1102–1111
14. Ghudumidze N, Chkonia I, Shapovalova N, Sadunishvili T, Meiphariani A (2006) Study of bacteriophages against the tomato with some bacterial etiologies. Proc Georgian Acad Sci Biol Ser B 4:21–24
15. Ghudumidze N, Shapovalova N, Giorkhelidze D, ZaaliShvili G, Sadunishvili T (2007) The morphological properties of phages specific for *Xanthomonas vesicatoria* tomato bacterial strains. Proc Georgian Acad Sci Biol Ser B 5:26–29
16. Ghudumidze N, Alavidze Z, Chkonia I, Eliashvili P, Giorgobiani N, Shapovalova N, Meiphariani A, Sadunishvili T (2007) Effective controlling of bacterial spot in tomato with bacteriophages. Proc Georgian Acad Sci Biol Ser B 5:8–11
17. Goldfarb DM (1961) Bacteriophagy. Med. Gaz, Moscow, p 295, In Russian
18. Gómez P, Buckling A (2011) Bacteria-phage antagonistic coevolution in soil. Science 332:106–109
19. Hert AP (2001) Relative importance of bacteriocin-like genes in antagonism of T3 to T1 strains of *Xanthomonas campestris* pv. *vesicatoria*. MS thesis, University of Florida, Gainesville
20. Iriarte FB, Balogh B, Momol MT, Smith LM, Wilson M, Jones JB (2007) Factors affecting survival of bacteriophage on tomato leaf surfaces. Appl Environ Microbiol 73(6):1704–1711
21. Jackson LE (1989) Bacteriophage prevention and control of harmful plant bacteria. US patent 4,828,999
22. Jones JB (1991) Bacterial spot. In: Jones JB et al (eds) Compendium of tomato diseases. APS Press, St. Paul, p 27
23. Jones JB, Bouzar H, Somodi GC, Stall RE, Pernezny K, El-Morsy G, Scott JW (1998) Evidence for the preemptive nature of tomato race 3 of *Xanthomonas campestris* pv. *vesicatoria* in Florida. Phytopathology 88:33–38
24. Jones JB, Jackson LE, Balogh B, Obradovich A, Iriarte FB, Momol T (2007) Bacteriophages for plant disease control. Annu Rev Phytopathol 45:245–262
25. Jones JB, Vallad GE, Iriarte FB, Obradovich A et al (2012) Considerations for using bacteriophages for plant disease control. Bacteriophage 2:208–214
26. Koller W (1998) Chemical approaches to managing plant pathogens. In: Ruberson JB (ed) Handbook of integrated pest management. Dekker, New York
27. Kutter E, Sulakvelidze A (2005) Bacteriophages:biology and applications. CRC Press, Boca Raton, 500p
28. Labrie SJ, Samson JE, Moineau S (2010) Bacteriophage resistance mechanisms. Nat Rev Microbiol 8:317–327
29. Leboeuf J, Cuppels D, Dick J, Pitblado R, Poewen St, Celetti M (2005) Bacterial diseases of tomato; Bacterial Spot, Bacterial Speck, Bacterial Canker. Queen's Printer for Ontario. Factsheet ISSN:1198-712X, 363–365
30. Louws EJ, Wilson M, Cambell HL, Cuppels DA, Jones JB, Shoemaker PB, Sahin F, Miller SA (2001) Field control of bacterial spot and bacterial speck of tomato using a plant activator. Plant Dis 85:481–488
31. Mao W, Lewis JA, Lumsden RD et al (1998) Crop protection. Crop Prot 17:535–542

32. McManus PS, Stockwell VO, Sundin GW, Jones AL (2002) Antibiotic use in plant agriculture. Annu Rev Phytopathol 40:443–465
33. Momol MT, Jones JB, Olson SM, Obradovic A, Balogh B, King P (2002) Integrated management of bacterial spot on tomato in Florida. Rep PP110, EDIS. Inst. Food Agric. Sci., Univ. FL
34. Moore ES (1926) D'Herelle's bacteriophage in relation to plant parasites. S Afr J Sci 23:306–310
35. Obradovich A, Jones J, Momol M et al (2004) Management of tomato bacterial spot in the field by foliar applications of bacteriophages and SAR inducers. Plant Dis 88:736–740
36. Pal KK, McSpadden G (2011) Biological control of plant pathogens. The Plant Health Instructor, 10, 1094/PHI-A-2006-1117-02:1–25
37. Sadunishvili T, Giorgobiani N, Amashukeli N et al (2012) Strategy of biological control of phytopathogenic bacteria in Georgia. Ann Agrar Sci 10:62–66
38. Sulakvelidze A, Barrow P (2004) Phage therapy in animals and agribusiness. Bacteriophages Biol Appl 335:380
39. Svircev AM, Lehman SM, Kim WS, Barszcz E et al. (2006) Control of the fire blight pathogen with bacteriophages. In: Zeller W, Ulrich C (eds) Proceedings of the 1st international symposium on biological control of bacterial plant diseases, 408: 259–261
40. Tanaka H, Negishi H, Maeda H (1990) Control of tobacco bacterial wilt by an avirulent strain of *Pseudomonas solanacearum* M4S and its bacteriophage. Ann Phytopathol Soc Jpn 56:243–244
41. Thayer PL, Stall RE (1961) A survey of *Xanthomonas vesicatoria* resistance to streptomycin. Proc Fla Hort Soc 75:163–165
42. Thomas RC (1935) A bacteriophage in relation to Stewart's disease of corn. Phytopathology 25:371–372
43. Wilson MS, Hirano S, Lindow SE (1999) Location and survival of leaf-associated bacteria in relation to pathogenicity and potential for growth within the leaf. Appl Environ Microbiol 65:1435–1443

Chapter 4
Isolation of Recombinant Antibodies That Recognize Native and Accessible Membrane Biomarkers

Ario de Marco

Abstract Monoclonal antibodies are indispensable reagents for diagnostics, but their isolation and production by means of conventional immunization and hybridoma technology is tedious, time-intensive, and expensive. Panning large pre-immune phage display libraries is a reliable and fast alternative; in addition, this approach presents the advantage that can be applied to toxic and non-immunogenic antigens. Furthermore, panning can be performed directly on whole cells to identify antibodies that will recognize their membrane-bound antigens in their native conformation and lipid environment. This opportunity is particularly meaningful when it is necessary to isolate antibodies that are able to bind accessible epitopes in vivo, as it is the case of biomarkers exposed at the cell membrane surface of pathogenic micro-organisms. Nevertheless, the effectiveness of biopanning can be undermined by different biases. In this presentation, the most common shortcomings of antibody phage display will be discussed, taking into account the most recently published reports as well as the experiments performed in our laboratory. Furthermore, some strategies useful to overcome these drawbacks will be illustrated and integrated with practical advice.

4.1 Introduction

Monoclonal antibodies are valuable and often essential reagents for diagnostics and therapy. Hybridoma technology [1] enables the ability to reproduce indefinitely single clones with defined characteristics; this opportunity allowed for the establishment of experimental references as well as for reliable data comparison among different labs. The successive most significant technological progress in the field of the clonal antibody isolation was the adaptation of the phage display technology [2]

A. de Marco (✉)
Department of Biomedical Sciences and Engineering, University of Nova Gorica, Glavni Trg 9, SI-5261 Vipava, Slovenia
e-mail: ario.demarco@ung.si

T.A. Camesano (ed.), *Nanotechnology to Aid Chemical and Biological Defense*,
NATO Science for Peace and Security Series A: Chemistry and Biology,
DOI 10.1007/978-94-017-7218-1_4

to generate collections of antibody fragments. Their availability enabled to pan in vitro against purified antigens to select specific binders [3]. Phage display libraries of antibodies in different formats such as Fab fragments, scFvs, and VHHs are now available and their specificities have been recently reviewed [4]. Yeast, ribosome, and bacterial display of antibodies and other alternative scaffolds represent variants that share with phage display the same possibility to link a phenotype (the binding polypeptide) to its corresponding genotype. In the present methodological contribution, the focus will be on the advantages and the shortcomings offered by a specific class of antibody libraries, namely the pre-immune collections of antibodies in single-domain (VHHs, known also as nanobodies) format.

The interest for both synthetic and naïve single-domain pre-immune libraries [5–10] is multiple: (i) they are prepared once and then can be used "forever" (thousands of aliquots are generated at the initial step) because they are unbiased with respect to any antigen; (ii) consequently, their use is progressively less expensive (no animal purchase and care, no cost for generating further libraries) and; (iii) extremely faster (10 days for performing three panning rounds instead of the several months necessary for animal immunization, new library preparation, and panning); (iv) they can also be used when the target molecule is not immunogenic or is toxic; (v) they can provide antibodies for unknown antigens specifically expressed in a particular context (markers of cell sub-populations, mutants); (vi) they are particularly suitable for identifying antibodies towards biomarkers present in unpurified biological material. Finally, in comparison to libraries of antibody fragments composed of two domains, single-domain libraries possess the further advantage that one single sequence must be cloned in the phagemid vector during their preparation [11].

Initially, it was doubted that antibodies that did not undergo somatic maturation could reach affinities sufficiently elevated for their antigens to make them biotechnologically useful. In the reality, large collections of recombinant antibodies provide a structural variability sufficiently elevated to compensate the lack of the in vivo adaptation process and to recover clones characterized by sufficient specificity and affinity for any challenged antigen [12]. Fab fragments with affinity in the low nM range were recovered from synthetic repertoires [13] and VHHs with pM affinities were isolated from a llama naïve library [10]. However, it remains controversial what level of variability must be present in a collection to assure its convenient usability. More and more, library evaluation is performed looking at their functional size rather than at their theoretical total diversity [4]. This parameter takes in account that many factors related to antibody folding and stability will influence and reduce the actual number of antibodies that are correctly exposed during panning and that will be easily produced after selection. A trivial example is offered by the comparison of VHH and scFv libraries. After cloning, any single VHH represents an independent binding domain that has the theoretical capacity to bind to a compatible epitope. In contrast, many of the randomly created VH/VL pairs cannot match because of steric hindrance and are not functional. Consequently, part of the assembled scFvs contributes to the theoretical total diversity of the library but not to its functional diversity. Furthermore, these non-functional antibodies are prone to aggregation and consequently their stickiness can contribute to increase the background represented by non-specific binders that contaminate the panning output. Their elevated number diminishes the percentage of functional binders specific for the target(s) isolated after each panning round, and renders the isolation of functional antibodies from the total pool more difficult.

The importance of improving the factors influencing the expression rate, the structural stability, and the binding features of the antibody fragments used for phage display [14] is therefore evident. In this perspective, the selection of heat-resistant domains has been used to identify robust scaffolds with low aggregation propensity inside a population with large sequence diversity [15]. Otherwise, rational design can be envisaged for scaffold development since, differently from libraries prepared using antibodies recovered from animals, the construction of synthetic libraries allows for sequence optimization. The contribution to structural stability and non-specific stickiness of key amino acids present at the different positions in single-domain antibody sequences of both *Camelidae* and human origin has been thoroughly analyzed, with the aim to define "golden rules" for the choice of the sequences corresponding to frameworks and complementarity determining regions [6, 9, 16–19]. The reduced mass and higher structural stability of (synthetic) VHHs explains also why they can be easily reconstituted into IgG-like molecules [20] without the drawback observed when operating with scFv synthetic libraries. In this last case, it has been observed that antibody fragments selected from completely synthetic libraries were more sensitive to lose specificity when reconstituted into IgG format than binders isolated from naïve libraries [21]. The authors suggested that natural repertoire would have a sort of evolutionary advantage in comparison to structures that have not been challenged for their stability in any organism. It would be interesting to compare the format-dependent stability of synthetic binders like adhirons that are monomeric in origin [22].

Pre-immune antibody libraries can be then used for selecting binders specific for any antigen. Of course, the quality of the selected antibodies will depend on the quality of the antigen and, in this perspective, the antigen purification process can result more critical than the procedure to recover the corresponding antibodies. However, although the standard selection approach based on phage display has been well established and is generally reliable, it may hide some dangers that will be described later in this text.

4.2 Reasons and Limits of Panning on Whole Cells

Antibodies are used for a large spectrum of different applications. Among them, the clinical exploitation of antibodies and their fragments covers a particularly important role and has a primary economic relevance. In most of the therapeutic and diagnostic applications, it is necessary that the antibodies can recognize their antigens in vivo. This condition is not always respected when conventional antibodies are used because they could have been raised against recombinant antigen samples that did not preserve their native folding. Consequently, immune-dominant epitopes that are not present or not accessible in the native macromolecule conformation can be exposed and catalyze the immune response, whereas useful epitopes are neglected. Some classes of antigens such as membrane proteins are particularly difficult to produce in recombinant form because they are often stabilized by disulfide bonds and hydrophobic interaction with lipids. Furthermore, they can undergo post-translational glycosylation modifications that are difficult to reproduce using bacteria, yeast, or insect cells as recombinant systems. These limitations result in antigens that can be unsuitable for obtaining reliable antibodies via both animal immunization and in vitro selection. The direct panning on cells, in contrast, is performed against native antigens stabilized by their natural lipid milieu, namely exactly the same structures that are expected to be accessible in vivo.

Panning on cells can be extremely effective to isolate binders specific for a cellular sub-population, or even for a single antigen differentially expressed among two cell lines. For this approach, collections have been successfully used that originated from both cell-immunized animals and large unbiased pre-immune libraries [23, 24]. The output quality depends on the efficacy of the depletion strategy envisaged to eliminate the antibodies that recognize epitopes present in both cell groups. The protocol illustrated in Fig. 4.1 shows that the antibody pool

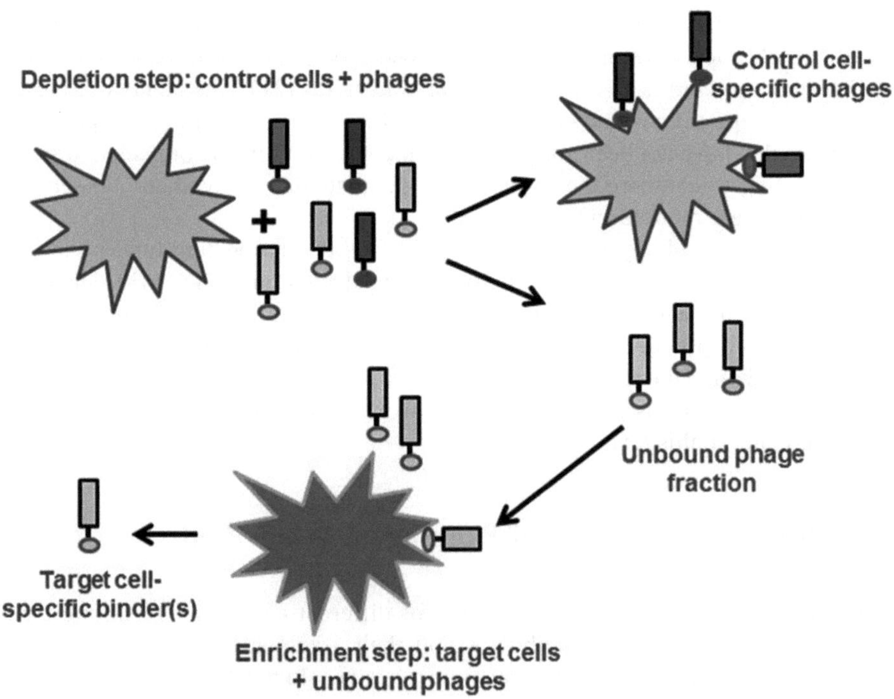

Fig. 4.1 Schematic succession of the panning on cells. Whole cells in combination with a pre-immune phage display library were effectively used to identify antibodies capable to discriminate between cell sub-populations

is initially challenged against the control cell population. It is important to get rid of both sticky binders and antibodies that recognize epitopes shared by the target population. For this reason, we have seen that working with a poorly redundant collection (for instance, 10^{11} phages for a library with a diversity of 10^9 clones, namely 100 replicates of each clone) and enchaining two independent depletion rounds in the presence of ten million cells each is very effective to eliminate irrelevant phages and to drastically reduce their background during screening. Once eliminated the bound antibodies, the unbound fraction is used to challenge the target cell population. Excellent results have been obtained with as few as 20,000 cells [25] and probably the limit could be further decreased when a single antigen – and consequently a relatively low number of different antibodies- discriminates between two cell populations. Present technology enables to isolate and manipulate single cells. Therefore, it would be imaginable to separate a small population of heterogeneous target cells and to perform elution on individual cells with the aim of collecting specific "fingerprinting" antibody pools.

Several groups have exploited phage display, panning on whole tumor cells to select peptides [26] and antibody fragments capable of recognizing "cancer biomarkers", namely antigens specifically (over)expressed at the surface of tumor cells [27–30]. Of course, the specificity of such antigens is always relative to the control cell populations used to deplete the antibody collection from the binders that recognize "healthy" epitopes, and must be validated by a compete screening on arrays of healthy and tumor cells/tissues. The largest systematic work performed to identify antibody/antigen couples specific for carcinomas [28] resulted in the characterization of 29 tumor associated antigens [31]. Apparently, the selection on whole cells has a bias for membrane proteins with large and glycosylated ectodomains such as the adhesion proteins of the Ig superfamily. This can be explained in terms of privileged accessibility of the phages to the most external displayed proteins and, in contrast, steric hindrance impairing the contact with proteins of smaller mass that are partially masked by the larger "competitors". Nevertheless, the system can be harnessed by overexpressing specific targets to change the protein ratio at the cell surface. Adopting this approach, it was possible to recover nanobodies specific for GPCR proteins with minimal extra-cellular domain (62 aa) such as CXCR4 [23].

In vivo, immunogenicity of specific polypeptide sequences inside the whole antigen is the driving force leading to the selection of pools of antibodies specific for relatively few epitopes. For instance, polyclonal antibodies apparently recognize no more than three to four epitopes of large antigens [32, 33]. Whereas immune dominance has been thoroughly investigated inside an immunological perspective, no attempt has been tried to understand why some sequences or tridimensional structures are extremely more active than others in stimulating the antigenic response independently upon their similarity with sequences present in the host organism. A sort of immune dominance is recognizable during panning in vitro performed with pre-immune libraries as well, although no immunological mechanism is really active during selection and, in contrast, it could be expected that antibodies would recognize different epitopes stochastically. Hoogenboom et al. [24] identified as many as 20 different antibodies when panning a pre-immune scFv library against CHO cells overexpressing human CD36 but remarked that all recognized a unique overlapping epitope. Similarly, panning on myoblasts expressing the rat CD36 homologue revealed similar selection dominance. Seven of the anti-(rat) CD36 cross-reacted with human CD36 and all the in vitro selected antibodies were inhibited by a monoclonal anti-CD36 known to recognize a functional epitope that is considered immunodominant in vivo. In this specific case, there is coincidence between immunodominant epitope in vivo (conserved between human and rat) and in vitro. This observation suggests the hypothesis that at least a component contributing to immune dominance in vivo is not related to the divergence between the sequence/structure features of the antigen and those of its homologue/other proteins in the host organism. It also means that in vitro immune dominance is not the mere bias resulting from the library design. The persistence of the same selective mechanism in vitro would rather indicate that specific structural conformations could be preferentially favored over others independently upon their conservation or uniqueness among different organisms. We have noticed a similar coincidence by panning on HER2 positive cells (SKBR3) after depletion on HER2 negative cells (MCF10A). Six antibodies with unique sequences and totally different CDRs, all specific for HER2, were characterized by epitope binning and all them resulted mutually competitive. Furthermore, they competed with the commercial monoclonal antibody trastuzumab as well. This was a further surprise since the recombinant antibodies used in this panning were single-domain antibodies (nanobodies/VHHs) [10]. This class of antibodies is characterized by a large CDR3 forming a protruding loop that binds preferentially to deep epitopes inaccessible to the flat paratope of conventional antibodies, a reason for which the two antibody classes are thought to recognize complimentary epitopes [34]. Nevertheless, some antigen structure conformations may result sufficiently thermodynamically appealing to accommodate binders with paratopes possessing extremely dissimilar structural features. This conformational advantage would confer an immune dominance recognized by both in vitro and in vivo selection mechanisms.

4.3 Harnessing the System to Isolate Antibodies Possessing Specific Functions

Panning on cells can also be exploited to lead the antibody selection towards epitopes of particular applicative interest [35]. Antibodies that recognize an epitope common to the mouse and human isoforms of the same antigen allow for the analysis of both clinical samples and murine models. Since most of the monoclonal antibodies are raised in mice, it is relatively rare to obtain hybridomas expressing suitable antibodies for such epitopes because they possess low immunogenicity in the injected animals. In contrast, cross-reacting antibodies can be successfully isolated in vitro by applying the scheme depicted in Fig. 4.2 and that we used to identify an anti-ALK antibody specific for both the human and the mouse

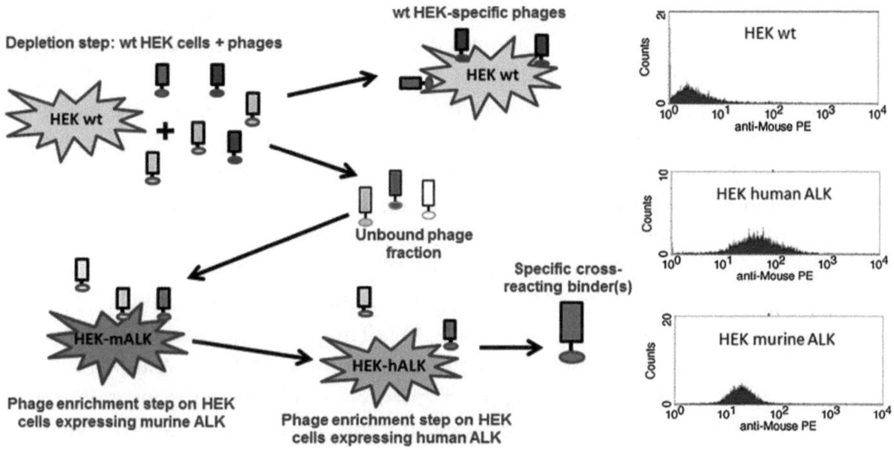

Fig. 4.2 Strategy for selecting cross-reacting antibodies. Wild type (wt) HEK 293 cells and the same cells overexpressing the recombinant ectodomain of mouse and human ALK at their surface have been used to select antibodies that recognized an epitope shared by the two protein isoforms

isoforms. The pre-immune library was initially depleted using wild type HEK293 cells and the unbound phage fraction enriched on HEK293 cells expressing mouse ALK at the surface. The positive clones were further challenged in the presence of HEK293 cells displaying human ALK. Three different clones that specifically reacted with both human- and mouse-expressing HEK293, but were negative on the wild type cells were isolated (Fig. 4.2) and their binding to ALK, were confirmed by ELISA using the recombinant protein ectodomain. A specular approach aimed at identifying antibodies specific for only one isoform/mutant of an antigen considers removing cross-reacting phages (pan-antibodies) by depleting the library in the presence of an excess of the undesired antigens before allowing the unbound phages to selectively bind to the target antigen [36].

Sometimes it is necessary to isolate separate antibodies that remain accessible at the cell surface from others that internalize rapidly. For instance, the first group can be envisaged for ADCC whereas the second for delivering antibody-drug conjugates. The simple shift of the temperature from 4 °C to 37 °C during the incubation of the antibody phage display libraries in the presence of the target cells is sufficient to induce the intracellular accumulation of binders associated to internalizing antigens [37]. Such phage particles can be selectively recovered from the cells after removal of the binders that remained bound to the external cell surface.

Antibodies can be isolated also according to their thermal stability. Since this biophysical character correlates with structural stability, introducing a heat denaturation step during panning enables to select antibodies with lower propensity to aggregation and higher resistance to physical and chemical destabilizing conditions [38]. Similarly, panning in vitro can be also used to select antibodies that show particularly high stability towards some peculiar conditions. For instance, the addition of high detergent concentrations or peptidases in the washing buffer will favor the recovery of clones that are resistant to these denaturing factors and that will be suitable for applications in the presence of such chemical and biological elements. This strategy has been exploited to isolate VHHs that remain functional in shampoos and animal gut [39, 40].

However, the most direct approach to identify antibodies possessing biological functions is to couple panning to competitive elution or epitope masking performed with already characterized effector molecules. In the case of the panning on whole cells, by such a way it is possible, for instance, to isolate antibodies that share the epitope with a receptor ligand and then to evaluate their potential agonist/antagonist effect. In the past we have recovered a single-domain antibody with antagonist effect on the FGFR1 by eluting the phage display antibodies bound to the receptor in the presence of the natural ligand FGF2 [41]. This antibody was able to prevent the FGF2-dependent activation and internalization of FGFR1. Although so far this panning protocol has been applied mostly to soluble antigens rather than to whole cells [42], there are no reasons for this limitation. This approach can also be used to identify recombinant antibodies that recognize the same epitope of already characterized conventional antibodies [43, 44]. The advantage is that recombinant antibody fragments, in contrast to their conventional homologues, can be easily engineered into structures with improved therapeutic potential such as biparatopic/bispecific/chelating antibodies and nanoparticles [45–47]. The examples reported in the literature show that such "functional antibodies" can be isolated from both epitopes with limited immune activity (the case of FGFR1 described in [41]) and others that are immune dominant, as in the case of CD36 [24] and HER2 (this paper).

4.4 Technical Aspects of Panning

Biopanning phage display libraries should be a straight forward procedure that enables to isolate the antibodies/binders with the highest affinity for their antigen by repeated selection rounds. Nevertheless, a number of recurrent drawbacks has been noticed in the praxis. Probably, it would be more appropriate to say that biopanning of phage display collections rewards the fittest clones rather than the most affine. In the next sections the factors that contribute to such overall fitness will be summarized.

4.4.1 The Contribution of Deep Sequencing Analyses to Understand the Selection Mechanisms of Phages

Phagemids are preferentially used as vectors for displaying polypeptides on the surface of filamentous phage M13 for two main reasons: (i) they enable monovalent display and (ii) they are more genetically stable, namely they have lower tendency than phage vectors to delete unnecessary DNA [48]. Monovalent display should avoid the contribution of avidity as a selection factor during biopanning. Therefore, exclusively the binder affinity for its antigen should lead to its enrichment during the repeated panning cycles. However, already the first scientists dealing with the phage display technique observed that the phage amplification step introduces a strong bias towards some sequences [49]. Statistical analyses suggested a relationship between the presence of specific amino acids at determined positions of the peptide sequence and the efficiency of its display [50]. Empiric investigations demonstrated that also mutations in the non-coding sequence could regulate phage replication and explain their differential accumulation [51]. The enrichment of target-unrelated phages (corresponding to so called "parasite" sequences) can therefore be attributed to propagation-related reasons rather than to selection-related factors and can lead to a clone diversity collapse in the presence of repeated amplification cycles [52, 53].

The anomalies of biopanning performed using phage display libraries have been mostly evidenced applying new generation sequencing to peptide collections. This approach allowed demonstrating that the library diversity is not reduced – as expected- during selection, but during the amplification steps [54–56]. As a consequence, it is not possible to infer a correlation between abundance of the clones identified in the screening step and their affinity because "faster" phages will always overrun slow-growing phages. Increased stringency during panning, shorter amplification time, and direct selection of released phages without purification might mitigate growth competition among binders that recognize the antigen with similar efficiency and would contribute to eliminate part of the fast-growing non-specific binders [57].

In contrast to what has been performed with peptide libraries, the analysis of antibody fragments obtained starting from conventional IgGs (Fabs or scFvs) presents a major shortcoming due to the limited length of the sequences that can be identified in a single run. Practically, it means that VHs and VLs are sequenced independently. Consequently, the repeats of one specific CDR3 can be monitored over the biopanning cycles, but no direct analysis of complete individual clones is feasible since the same CDR3 could be present in a domain coupled to different partners [58]. This limitation is overcome when using single-domain antibodies because their whole sequence can be identified in a single run. Therefore, we applied deep-sequencing to evaluate whether the bias observed during biopanning of phage display libraries of peptides and, probably, scFvs [59, 60] were conserved during panning of nanobody libraries. We depleted a large synthetic phage display library using the MCF10A cell line (HER2 neg) and challenged the unbound fraction in the presence of the HER2 positive cells SKBR3. The clones positive for SKBR3 and negative for MCF10A in flow-cytometry were further screened by ELISA using the purified recombinant ectodomain of the HER2 receptor. We did not sequence negative clones searching for parasite phages but analyzed the frequency of the HER2-positive binders isolated after biopanning in the second and the third selection round. In parallel, the affinity of six different nanobodies with apparent diverse binding capacity according to the ELISA and FACS data was measured by Surface Plasmon Resonance. The results summarized in Table 4.1 confirm that affinity is not the sole selection driver during biopanning. For instance, the antibodies a and c are equally represented after the second panning round, but

Table 4.1 Deep sequencing demonstrates that affinity is not the unique driver during antibody selection by phage display. The number of clones corresponding to six different antibodies selected during the same panning procedure were measured in the output phage population issued from the rounds 2 and 3 and their affinity for their substrate human HER2 was calculated by SPR

		Sequence repeats	
Antibodies	K_{D} nM	Output round 2	Output round 3
A	2	317	18,786
B	20	1,738	372,729
C	400	367	79,640
D	900	1,949	162,148
E	680	1,039	58,909
F	4	275	20,332

c resulted amplified more efficiently at the end of the third round than a (215 times instead of 60, respectively) although its affinity for the antigen was 200 times lower (400 and 2 nM, respectively). The second most represented clone after the third biopanning round was still the less affine binder of the group (d, 900 nM). Among the six nanobodies, only b seems being selected according to both affinity (20 nM) and amplification propensity.

4.4.2 Phage Multi-infection and Other Anomalies

The theory postulates that single bacteria can be infected by only one plasmid. In the past, working with nanobodies recovered from a naïve library, we often had troubles to get complete sequences of our selected binders. This shortcoming is common to people working with immune libraries as well [61] and researchers usually attribute this fact to the scarce quality of the DNA used for sequencing. Synthetic libraries in which only the CDRs are hypermutated but the frameworks are conserved allow for a systematic analysis of the analyzed sequences. The electropherogram reported in Fig. 4.3 (top) shows the CDR1 region flanked by frameworks 1 and 2. The framework sequences are unambiguous whereas two conflicting sequences clearly appear in the CDR1 region. This result can be interpreted as the consequence of a double infection of a single bacterium. In the praxis, this situation is annoying because impairs the recovery of positive clones and the determination of their sequence. Sometimes, iterative dilution and growth cycles are necessary to discriminate the two clones (Fig. 4.3, bottom) and this tedious procedure is amplified when several sequences indicate multiple infections. We have found that this drawback can be prevented by preparing a phagemid mini-prep using the phage output obtained at the end of a biopanning cycle and using this DNA for transforming bacteria. This procedure assures the growth of colonies with no detectable double infection because apparently overcomes the critical step represented by multiple bacterial infection by means of phages. Summarizing,

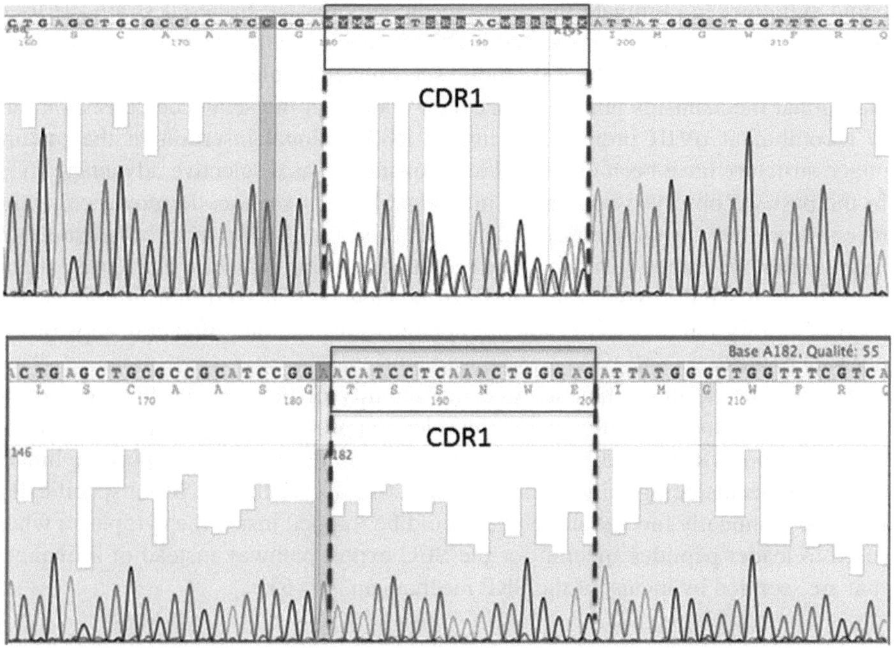

Fig. 4.3 Identification of phage double infection. Synthetic libraries with constant framework and hypervariable CDRs allow for easy identification of multiple phage infection of host bacteria. When two independent phagemids are purified and sequenced together, the signal of the conserved bases is amplified whereas it is conflicting in the variable CDR regions (in this example, CDR1). Clonality is recovered after re-isolation

although phages use an infection mechanism compatible with multiple infection of a single bacterial cell [62], multiple infection by phagemids is extremely rare.

Further anomalies during biopanning with phage display libraries derive from the fact that the use of phagemids does not completely prevent the presence of multivalent display. The consequent increased avidity will favor phage particles that have higher propensity to expose multiple binders. We noticed another specificity of phagemid-based libraries. As recalled at the beginning of this chapter, phagemids should be less prone to eliminate unnecessary DNA than phages [48]. However, we identified a relatively high number of unrelated sequences that shared the same conserved mutation among the binders selected during biopanning against different classes of antigens. This mutation transformed the original glutamine codon CAG into the amber stop codon TAG. Since we used the suppressor strain TG1 during both panning and the expression of the phages for the preliminary screening, such mutation remained ineffective. Nevertheless, this observation could suggest that phagemids undergo mutagenesis to eliminate inserted sequences and that this process can have actually eliminated several valuable clones. We appreciate the presence of this process only because one of the mutations is detectable without being sufficient to eliminate the clone in the suppressive bacterial strain we used for phage expression. A unexpected high frequency of amber stops was identified also in antibodies isolated by a synthetic scFv phage display library [21]. Different mutational mechanisms that introduce stop codons in the sequence corresponding to recombinant pVIII protein and impair its functional insertion in the mature phage structure have been described and considered as a selective advantage [63]. In the past we observed apparently inexplicable facts such as the presence of the recognition sites for the restriction enzymes used for cloning the whole nanobody insert within the cloned sequence. The presence of a high mutation rate during phage amplification would explain the onset of sequences that were not present in the original library but that were rather generated along the amplification steps of the panning process.

Finally, the folding efficiency and the stability of the exposed binders can be also relevant parameters influencing the selection process during biopanning. In the case of both DARPins and VHHs, the secretion route selected to export the binder molecules seems being critical for their effective accumulation. More specifically, thermodynamically favored molecules could be trapped inside the cytoplasm when fused to leader peptides specific for the SEC export pathway instead of sequences that are secreted by means of the SRP mechanism [64, 65].

4.4.3 Combining Phage and Yeast Display

As we have seen in the previous chapters, repeated biopanning cycles seem to favor clones that are amplified faster, and deep sequencing analyses have shown that they are over-represented even when they have worse binding capacity. Generally, 10^6 phage particles are recovered as the output of the second/third round of biopanning and this number can increase in the successive cycles. At the same time, conventional screening methods based on FACS or ELISA allow for the analysis of only few hundreds of clones. In these conditions, the statistic chances to identify good binders that are amplified slowly diminish progressively after each amplification cycle. This observation led to the development of protocols aimed at limiting the amplification cycles by using deep sequencing for direct recovery of potential specific binders or to perform amplification in conditions that prevent biased selections [52, 56, 59, 60]. An alternative method consists in combining phage and yeast display (Fig. 4.4, [66]). The first format allows for preparing phage display collections characterized by elevated variability (10^9–10^{10} different clones) to select as many as possible specific binders. Mathematical models foresee that using libraries of such diversity 10^2–10^5 different sequences with comparable affinity and sufficient specificity should be recovered for any given antigen [54]. However, phage display is not a suitable format for high-throughput screening of the clones selected by panning and in most of the cases no more than few hundreds of clones are analyzed by ELISA or FACS, despite the output of the first/second panning rounds counts 10^5–10^7 phages. Therefore, diversity collapse

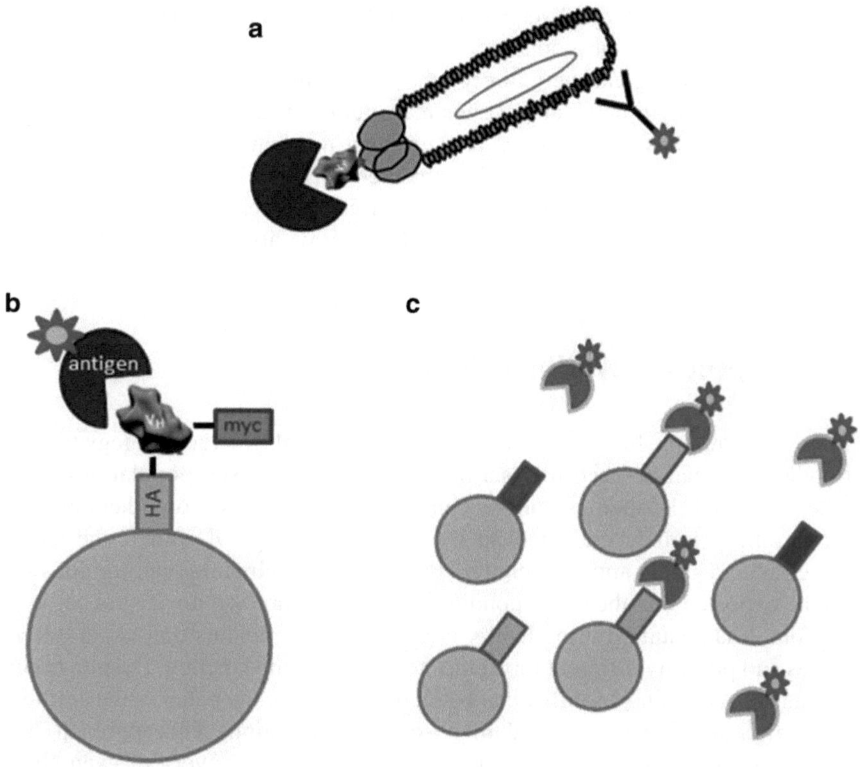

Fig. 4.4 Complementarity of phage and yeast display. Phage and yeast display exploit complementary principles for the isolation of antigen-specific antibodies: (**a**) the antibody exposed on a phage binds to an immobilized antigen and the binding is visualized by means of an anti-phage antibody; (**b**) the expression of antibodies exposed on a yeast cell can be evaluated by targeting the tags present at the C- and N-term of the antibody fragment. Such antibody can bind to a soluble labeled antigen; (**c**) the yeast/antigen complex can be visualized by direct or indirect antigen labeling and sorted by flow cytometry

due to amplification biases is exacerbated by the practical possibility to evaluate only a minimal percentage of the total phage population issued from panning. Yeast display can partially compensate these shortcomings. First, yeast uses a different secretion mechanism with respect to prokaryotic organisms. Consequently, phages that overgrow in bacteria might find unfavorable conditions in eukaryotic. Of course, specific sequence-related biases exist also for the yeast secretion system. So far, no report dedicated to understand how this "double bias" could influence the total library diversity has been published. Nevertheless, a sure advantage of alternating phage with yeast display is the fact that yeast cells can be sorted by flow-cytometry and, therefore, the whole antibody population displayed on yeast cells can be evaluated. The main limit of yeast is its low transformation capacity (max 10^7), but this number is compatible with the output of the first panning round of phage display. Therefore, the large diversity compatible with phage display libraries (10^{10}) is exploited to recover sub-populations of selected clones (10^5–10^6) that after the first round of panning can be entirely recombined into yeast display vectors and evaluated by flow cytometry, at least when a soluble antigen is available.

This approach has proved to be very efficient to identify a panel of antibodies with a far greater variety than usually obtained carrying out phage display alone [67, 68]. Furthermore, yeast display can be used to assess antigen specificity, to class their affinity, and to perform epitope binning without the necessity to purify the antibodies [69].

4.5 Perspectives: Antibody Biomarkers

Panning pre-immune libraries on whole cells is a straightforward strategy to isolate antibodies that can be used to discriminate between cell sub-populations. In this perspective, such antibodies can be considered themselves biomarkers and their practical utility is independent from the antigen identity. Of course, the labeling obtained with a single antibody confers minimal discrimination opportunities since it can be expected that the corresponding epitope is expressed on several cell types. Therefore, it is meaningful to identify combinations of binders that target different epitopes and possibly different molecules to use for multi-labeling. Despite immunodominance, a certain variability can be recovered inside the panel of the antibodies isolated after panning (Crepin et al. manuscript in prepration). This approach should increase specificity and allow for the identification and recovery of even slightly different cell types. The specific binding of a set of independent antibodies on a cell can serve as fingerprinting and be useful to gate and isolate specific sub-population of cells by flow-cytometry. Even though the literature reports mainly panning events performed on eukaryotic cells [27–30] and sometimes on pathogenic bacteria-like structures [70, 71], it can be envisaged to apply this protocol to structures as different as toxic microalgae and exosomes.

Acknowledgments The author wishes to thank Aurelie Schneider for her assistance, Creative Core AHA-MOMENT grant from Slovene Ministry of Economic Development and Technology as well as the European Fund for Regional Development – Cross-Border Cooperation Programme Italy-Slovenia 2007–2013, (Project PROTEO, Code N. CB166) for having supported this work with research funds.

References

1. Köhler G, Milstein C (1975) Continuous cultures of fused cells secreting antibody of predefined specificity. Nature 256:495–497
2. Smith GP (1985) Filamentous fusion phage: novel expression vectors that display cloned antigens on the virion surface. Science 228:1315–1317

3. McCafferty J, Griffiths AD, Winter G, Chiswell DJ (1990) Phage antibodies: filamentous phage displaying antibody variable domains. Nature 348:552–554
4. Ponsel D, Neugebauer J, Ladetzki-Bachs K, Tissot K (2011) High affinity, developability and functional size: the Holy Grail of combinatorial antibody library generation. Molecules 16:3675–3700
5. Arbabi- Ghahroudi M, MacKenzie R, Tanha J (2009) Selection of non-aggregating VH binders from synthetic VH-phage display libraries. Methods Mol Biol 525:187–216
6. Chen W, Zhu Z, Xiao X, Dimitrov DS (2009) Construction of a human antibody domain (VH) library. Methods Mol Biol 525:81–99
7. Koide A, Tereshko V, Uysal S et al (2007) Exploring the capacity of minimalist protein interfaces: interface energetics and affinity maturation to picomolar KD of a single-domain antibody with flat paratope. J Mol Biol 373:941–953
8. Liu JL, Anderson GP, Goldman ER (2007) Isolation of anti-toxin single domain antibodies from a semi-synthetic spiny dogfish shark display library. BMC Biotechnol 7:78
9. Mandrup OA, Friis NA, Lykkemark S, Just J, Kristensen P (2013) A novel heavy domain antibody library with functionally optimized complementarity determining regions. PLoS ONE 8:e76834
10. Monegal A, Ami D, Martinelli C et al (2009) Immunological applications of single domain llama recombinant antibodies isolated from a naïve library. Protein Eng Des Sel 22:273–280
11. Olichon A, de Marco A (2012) Preparation of a naïve library of *Camelidae* single domain antibodies: advantages and limitations of single-pot strategies. Methods Mol Biol 911:65–78
12. Perelson AS, Oster GF (1979) Theoretical studies of clonal selection: minimal antibody repertoire size and reliability of self-non-self discrimination. J Theor Biol 81:645–670
13. Griffiths AD, Williams SC, Hartley O et al (1994) Isolation of high-affinity human antibodies directly from large synthetic repertoires. EMBO J 13:3245–3260
14. Ewert S, Huber T, Honegger A, Plückthun A (2003) Biophysical properties of human antibody variable domains. J Mol Biol 325:531–553
15. Christ D, Famm K, Winter G (2007) Repertoires of aggregation-resistant human antibody domains. Protein Eng Des Sel 20:413–416
16. Lee CC, Perchiacca JM, Tessier PM (2013) Towards aggregation-resistant antibody by design. Trends Biotechnol 31:612–620
17. Ma X, Barthelemy PA, Rouge L, Wiesmann C, Sidhu SS (2013) Design of synthetic autonomous VH domain libraries and structural analysis of a VH domain bound to vascular endothelial growth factor. J Mol Biol 425:2247–2259
18. Saerens D, Pellis M, Loris R et al (2005) Identification of a universal VHH framework to graft non-canonical antigen-binding loops of camel single-domain antibodies. J Mol Biol 352:597–607
19. Zabetakis D, Anderson GP, Bayya N, Goldman ER (2013) Contribution of the complementary determining regions to the thermal stability of a single-domain antibody. PLoS ONE 8:e77678
20. Djender S, Schneider A, Beugnet A et al (2014) Bacterial cytoplasm as an effective cell compartment for producing functional VHH-based affinity reagents and *Camelidae* IgG-like recombinant antibodies. Microb Cell Fact 13:140
21. Chan CEZ, Chan AHY, Lim APC, Hanson BJ (2011) Comparison of the efficiency of antibody selection from semi-synthetic scFv and non-immune Fab phage display libraries against protein targets for rapid development of diagnostic immunoassays. J Immunol Methods 373:79–88
22. Tiede C, Tang AA, Deacon SE et al (2014) Adhiron: a stable and versatile peptide display scaffold for molecular recognition applications. Protein Eng Des Sel 27:145–155
23. Even-Desrumeaux K, Nevoltris D, Lavaut MN et al (2014) Masked selection: a straightforward and flexible approach for the selection of binders for specific epitopes and differentially expressed proteins by phage display. Mol Cell Proteomics 13:653–665
24. Hoogenboom HR, Lutgerink JT, Pelsers MM et al (1999) Selection-dominant and nonaccessible epitopes on cell surface receptors revealed by cell-panning with a large phage antibody library. Eur J Biochem 260:774–784

25. Shukla GS, Krag DN (2005) Phage display selection for cell-specific ligands: development of a screening procedure suitable for small tumor specimens. J Drug Target 13:7–18
26. Belizaire AK, Tchistiakova L, St-Pierre Y, Alakhov V (2003) Identification of a murine ICAM-1-specific peptide by subtractive phage library selection on cells. Biochem Biophys Res Commun 309:625–630
27. Geuijen CA, Bijl N, Smit RC et al (2005) A proteomic approach to tumour target identification using phage display, affinity purification and mass spectrometry. Eur J Cancer 41:178–187
28. Kurosawa G, Akahori Y, Morita M et al (2008) Comprehensive screening for antigens overexpressed on carcinomas via isolation of human mAbs that may be therapeutic. Proc Natl Acad Sci U S A 105:7287–7292
29. Poul MA, Becerril B, Nielsen UB, Morisson P, Marks JD (2000) Selection of tumor specific internalizing human antibodies from phage libraries. J Mol Biol 301:1149–1161
30. Ridgway JBB, Ng E, Kern JA et al (1999) Identification of a human anti-CD55 single-chain Fv by subtractive panning of a phage library using tumor and non-tumor cell lines. Cancer Res 59:2718–2723
31. Kurosawa G, Sumitomo M, Ukai Y et al (2011) Selection and analysis of anti-cancer antibodies for cancer therapy obtained from antibody phage library. Cancer Sci 102:175–181
32. Rockberg J, Löfblom J, Hjelm B, Uhlén M, Ståhl S (2008) Epitope mapping of antibodies using bacterial surface display. Nat Methods 5:1039–1045
33. Hudson P, Hudson EP, Uhlen M, Rockberg J (2012) Multiplex epitope mapping using bacterial surface display reveals both linear and conformational epitopes. Sci Rep 2:706
34. De Genst E, Silence K, Decanniere K et al (2006) Molecular basis for the preferential cleft recognition by dromedary heavy-chain antibodies. Proc Natl Acad Sci U S A 103:4586–4591
35. Hagay Y, Lahav J, Levanon A, Panet A (2003) Function-modulating human monoclonal antibodies against platelet-membrane receptors isolated from a phage-display library. J Thromb Haemost 1:1829–1836
36. Abe Y, Yoshikawa T, Inoue M et al (2011) Fine tuning of receptor-selectivity for tumor necrosis factor-α using a phage display system with one-step competitive panning. Biomaterials 32:5498–5504
37. Fransson J, Borrebaeck CA (2009) Selection and characterization of antibodies from phage display libraries against internalizing membrane antigens. Methods Mol Biol 480:113–127
38. Jespers L, Schon O, Famm K, Winter G (2004) Aggregation-resistant domain antibodies selected on phage by heat denaturation. Nat Biotechnol 22:1161–1165
39. Dolk E, van der Vaart M, Hulsik DL et al (2005) Isolation of llama antibody fragments for prevention of dandruff by phage display in shampoo. Appl Environ Microbiol 71:442–450
40. Harmsen MM, van Solt CB, van Zijederveld-van Bemmel AM, Niewold TA, van Zijederveld FG (2006) Selection and optimization of proteolytically stable llama single-domain antibody fragments for oral immunotherapy. Appl Microbiol Biotechnol 72:544–551
41. Veggiani G, Ossolengo G, Aliprandi M, Cavallaro U, de Marco A (2011) Single-domain antibodies that compete with the natural ligand fibroblast growth factor block the internalization of the fibroblast growth factor receptor1. Biochem Biophys Res Commun 408:692–696
42. de Marco A (2013) Methodologies for the isolation of alternative binders with improved clinical potentiality over conventional antibodies. Crit Rev Biotech 33:40–48
43. Forsman A, Beirnaert E, Aasa-Chapman MMI et al (2008) Llama antibody fragments with cross-subtype human immuodeficency virus type 1 (HIV-1)-neutralizing properties and high affinity for HIV-1 gp120. J Virol 82:12069–12081
44. Gruszka A, Martinelli C, Sparacio E, Pelicci PG, de Marco A (2012) The concurrent use of N- and C-terminal antibodies anti-nucleophosmin 1 in immunofluorescence experiments allows for precise assessment of its subcellular localisation in acute myeloid leukaemia patients. Leukemia 26:159–163
45. Oliveira S, Schiffelers RM, van der Veeken J et al (2010) Downregulation of EGFR by a novel multivalent nanobody-liposome platform. J Control Release 145:165–175
46. Roovers RC, Vosjan MJ, Laeremans T et al (2011) A bi-paratopic anti-EGFR nanobody efficiently inhibits solid tumor growth. Int J Cancer 129:2013–2024. doi:10.1002/ijc.26145

47. Villa A, Lovato V, Bujak E et al (2011) A novel synthetic naïve human antibody library allows the isolation of antibodies against a new epitope of oncofetal fibronectin. MAbs 3:264–272
48. Chasteen L, Ayriss J, Pavlik P, Bradbury ARM (2006) Eliminating helper phage from phage display. Nucleic Acids Res 34:e145
49. Peters EA, Schatz PJ, Johnson SS, Dower WJ (1994) Membrane insertion defects caused by positive charges in the early mature region of protein pIII of filamentous phage fd can be corrected by prlA suppressor. J Bacteriol 176:4296–4305
50. Rodi DJ, Soares AS, Makowski L (2002) Quantitative assessment of peptide sequence diversity in M13 combinatorial peptide phage display libraries. J Mol Biol 322:1039–1052
51. Brammer LA, Bolduc B, Kass JL et al (2008) A target-unrelated peptide in an M13 phage display library traced to an advantageous mutation in the gene II ribosome-binding site. Anal Biochem 373:88–98
52. Matochko WL, Cory Li S, Tang SK, Derda R (2014) Prospective identification of parasitic sequences in phage display screens. Nucleic Acids Res 42:1784–1798
53. Thomas WD, Golomb M, Smith GP (2010) Corruption of phage display libraries by target-unrelated clones: diagnosis and countermeasures. Anal Biochem 407:237–240
54. Derda R, Tang SK, Li SC et al (2011) Diversity of phage-displayed libraries of peptides during panning and amplification. Molecules 16:1776–1803
55. Matochko WL, Chu K, Jin B et al (2012) Deep sequencing analysis of phage libraries using Illumina platform. Methods 58:47–55
56. ‘t Hoen PAC, Jirka SM, Ten Broeke BR (2012) Phage display screening without repetitious selection rounds. Anal Biochem 421:622–631
57. Yu X, Barmina O, Burgoon M, Gilden D (2009) Identification of measles virus epitopes using an ultra-fast method of panning phage-displayed random peptide libraries. J Virol Methods 156:169–173
58. Ravn U, Gueneau F, Baerlocher L et al (2010) By-passing in vitro screening – next generation sequencing technologies applied to antibody display and *in silico* candidate selection. Nucleic Acids Res 38, e193
59. Ravn U, Didelot G, Venet S et al (2013) Deep sequencing of phage display libraries to support antibody discovery. Methods 60:99–110
60. Zhang H, Torkamani A, Jones TM et al (2011) Phenotype-information-phenotype cycle for deconvolution of combinatorial antibody libraries selected against complex systems. Proc Natl Acad Sci U S A 108:13456–13461
61. Kastelic D, Frković-Grazio S, Baty D et al (2009) A single-step procedure of recombinant library construction for the selection of efficiently produced llama VH binders directed against cancer markers. J Immunol Methods 350:54–62
62. Zeng L, Skinner SO, Zong C, Sippy J, Feiss M, Golding I (2010) Decision making at a subcellular level determines the outcome of bacteriophage infection. Cell 141:682–691
63. Ryvkin A, Ashkenazy H, Smelyanski L et al (2012) Deep panning: steps towards probing the IgOme. PLoS ONE 7:e41469
64. Monegal A, Olichon A, Bery N et al (2012) Single heavy chain antibodies with VH hallmarks are positively selected during panning of llama (*Lama glama*) naïve libraries. Develop Comp Immunol 36:150–156
65. Steiner D, Forrer P, Stumpp MT, Plückthun A (2006) Signal sequences directing cotranslational translocation expand the range of proteins amenable for phage display. Nat Biotechnol 24:823–831
66. Boder ET, Wittrup KD (1997) Yeast surface display for screening combinatorial polypeptide libraries. Nat Biotechnol 15:553–557
67. Ferrara F, Naranjo LA, Kumar S et al (2012) Using phage and yeast display to select hundreds of monoclonal antibodies: application to antigen 85, a tuberculosis biomarker. PLoS ONE 7:e49535
68. Ferrara F, D’Angelo S, Gaiotto T et al (2015) Recombinant renewable polyclonal antibodies. MAbs 7:32–41

69. Gai SA, Wittrup KD (2007) Yeast surface display for protein engineering and characterization. Curr Opin Struct Biol 17:467–473
70. Beer M, Liu CQ (2012) Panning of a phage display library against a synthetic capsule for peptide ligands that bind to the native capsule of *Bacillus anthracis*. PLoS ONE 7:e45472
71. Paoli GC, Chen CY, Brewster JD (2004) Single-chain Fv antibody with specificity for *Listeria monocytogenes*. J Immunol Methods 289:147–155

Chapter 5
Effect of Concentration on the Interactions of Gold Nanoparticles with Model Cell Membranes: A QCM-D Study

Elaheh Kamaloo, Christina Bailey, and Terri A. Camesano

Abstract The increasing applications of nanotechnology in everyday life require consideration of their interactions with living cells. There are different physical and chemical properties affecting the interactions of nanoparticles with cells, such as NP size, nature of functionalization or stabilizing groups, concentration, and the environment in which nanoparticles are interacting with cells. In this review, we summarize some of our previous studies done on the interactions of gold nanoparticles with supported lipid bilayers (SLB; models for cell membranes). These studies have been done via Quartz Crystal Microbalance with Dissipation (QCM-D) and they include NPs ranging in size from 2 to 40 nm at several concentrations. Their interactions with a SLB composed of L-α-phosphatidylcholine were characterized. In order to better understand how NPs behave in the environment, these interactions were also studied in the presence of different types of natural organic matter (NOM), including Aldrich humic acid, Suwannee River humic acid standard, Suwannee River fulvic acid standard, and Elliot soil humic acid. Here we review our previous findings while focusing on an example of the effect of concentration on NP-SLB interaction.

5.1 Introduction

Today, nanotechnology has a large impact on the way we live, as the presence of nanoparticles in commercially available products is becoming more common. There are many applications known for nanomaterials in medicine, catalysis, energy, sensors, and environment [1]. However, more information is needed on how nanoparticles interact with cells. Studies have been conducted to understand the effects of nanomaterials on humans and organisms, as well as in the environment

E. Kamaloo • C. Bailey • T.A. Camesano (✉)
Department of Chemical Engineering, Worcester Polytechnic Institute, 100 Institute Rd, Worcester, MA 01609, USA
e-mail: terric@wpi.edu

T.A. Camesano (ed.), *Nanotechnology to Aid Chemical and Biological Defense*, NATO Science for Peace and Security Series A: Chemistry and Biology,
DOI 10.1007/978-94-017-7218-1_5

[2, 3]. Different physical and chemical properties of nanoparticles such as size, concentration, and functionalization, can affect their interactions with biological systems. The effects of different nanoparticle sizes, functionalities, concentrations and the media in which nanoparticles interact with SLB have been characterized. In this study, we review our previous findings while focusing on an example of the effect of gold nanoparticle (NP) concentration on interactions with a model cell membrane.

A better understanding of how nanomaterials interact with cells is crucial to predict whether and how nanomaterials affect human health as well as how they impact on our environment by exercising any toxicity. There are different mechanisms of cytotoxicity such as disruption of cell membrane integrity, cell damage by generation of reactive oxygen species (ROS), damage to DNA, damage to the functionality of cellular proteins/enzymes, triggering of inflammation, oxidative stress, damage to mitochondrial function, etc. [4, 5]. All of these mechanisms may be due to atomic effects except for the disruption of cell membrane integrity, which is solely due to nanoparticulate effects, such as the size. For this reason, we have chosen to focus on the mechanism of cell membrane disruption using gold NPs. Gold nanoparticles have been given a lot of attention due to their wide range of applications in drug gene delivery, body imaging, and cancer therapy [6]. In addition, gold is inert in its molecular form. Therefore any cytotoxicity effect of gold nanoparticles originates from the nanoparticulate nature of the material. Size-dependent cytotoxicity of gold nanoparticles has been studied previously by cell culture assay [7]. The gold cluster size was varied from 0.8 to 15 nm. Particles 1–2 nm in size showed high toxicity while smaller particles and larger 15 nm particles were comparatively nontoxic. In another study, Mirovana et al. investigated the effect of size, concentration and exposure time of 13 and 45 nm gold NPs on human dermal fibroblasts. It was shown that for both particle sizes the cell doubling time was increasing with higher gold NPs concentration and longer exposure. In addition, by comparing the rate of apoptosis, it was found that there was a higher rate of apoptosis with either longer exposure or higher particle concentration [8].

In addition to chemical and physical properties of the nanoparticles, the environment in which the nanoparticles interact with SLB plays an important role. In order to stimulate the natural environment conditions, nanoparticles can be dispersed in natural organic matter, such as humic or fulvic acids. We have previously used a commercially-available humic acid (HA) as the representative of soil and aquatic environment. When nanoparticles disperse in a humic substance, their surfaces become covered by large organic molecules that change the stability and aggregation state of the nanoparticles [9]. Consequently this condition might change the toxicity of the nanoparticles.

Capping the surface of gold NPs with different surface charges using organic ligands has also proven to alter nanoparticle-membrane interactions. We have previously studied the role of four different capping agents, 1-propanethiol, 2-mercaptoethanol, 2-aminoethanethiol, and 3-mercaptopropionic acid in the presence of water and humic acid and compared these interactions to a citric acid-stabilized gold NP. Previous literature has shown that the surface charges and the end terminal group both have some degree of impact on the interaction when functionalized, which were consisted with our results when in a soil-like environment. However, in a water condition, all particles behaved similarly [10–12]. In order to study the interactions of nanoparticles with a model cell membrane, we have used Quartz Crystal Microbalance with Dissipation technique.

Quartz Crystal Microbalance with Dissipation (QCM-D) is a powerful acoustic system that provides information about mass change and viscoelastic properties of a film. In this instrument applying a voltage makes a quartz crystal oscillate and then the frequency of the oscillation and dissipation of energy of the system is measured. The change in frequency is related to mass change of the system and the dissipation of energy is primarily related to the viscoelasticity of the surface. In this study, QCM-D was used to monitor the mass and conformational change of a supported lipid bilayer formed on SiO_2 crystal. Many studies used QCM-D as a reliable method to confirm bilayer formation on the surface [13]. QCM-D has also been used to study the interactions between nanosized particles and supported lipid bilayers. For instance, McCubbin et al. used QCM-D technique for investigation of five different antimicrobial peptides (AMPs) to characterize the mechanism of interactions of them with SLBs [14].

5.2 Materials and Methods

5.2.1 Gold Nanoparticles

2 nm spherical gold nanoparticles were purchased from NANOCS (New York, NY) with a standard size distribution of 20 % measured by Dynamic Light Scattering (DLS) and Transmission Electron Microscopy (TEM) by the manufacturer. The received manufacturer concentration was 1.8×10^{14}, and 2 nm gold NP solutions were diluted to concentrations of 5.0×10^{10}, 1.0×10^{12}, and 1.0×10^{14} particles/mL to vary the concentrations by two orders of magnitude with each set. The zeta potential for these particles was measured by the manufacturer using DLS as −48.8 mV. All experiments were conducted with a bulk solution in de-ionized water, which was the original dispersant of the stock NP solutions. The stock NPs were stabilized with citrate and small amounts of tannic acid, stored at 7 °C in a light impenetrable container.

5.2.2 Vesicle Preparation

L-α-phosphatidylcholine (egg, chicken) (PC) was purchased from Avanti Polar Lipids with purity >99 % and prepared according to published procedures and similar to previous work [15, 16]. Unless otherwise stated, all other chemicals were purchased from Sigma-Aldrich (St. Louis, MO). In order to obtain a 100 mg/mL solution, 1 g of PC was added to 10 mL of ethanol and was then stored at −20 °C.

A stock PC vesicle solution was then made for storage up to 1 month. 0.15 mL of 100 mg/mL PC solution was dried with nitrogen gas and stored in a desiccator for 24 h before rehydration with 6 mL of Tris NaCl buffer solution. The Tris NaCl buffer was made at a pH of 7.8 of 10 nM Tri (hydrozymethyl) aminomethane with ≥99.9 % purity and 100 nM NaCl, +80 mesh particle size. The solution was first vortexed on a mini-vortexer for 15 s (Fisher Scientific Inc., Pittsburgh, PA) prior to five freeze-thaw-vortex cycles. Small unilamellar lipid vesicles were formed by sonicating the PC solution for 30 min in pulse mode with a 30 % duty cycle using an ultrasonic dismembrator (Model 150T, Fisher Scientific, Waltham, MA). The PC solution and newly formed small unilamlellar lipid vesicles were then centrifuged (Eppendorf Centrifuge 5415 D) for 10 min at 16,000× g at 23 °C in order to remove sonication probe particles from the solution [15, 16]. The supernatant was decanted from the pellet and stored under nitrogen gas at 7 °C. Before use, the PC solution was vortexed and diluted to 0.1 mg/mL in Tris-NaCl buffer.

5.2.3 Quartz Crystal Microbalance with Dissipation (QCM-D) and Sauerbrey Relationship

QCM-D was used to measure frequency (mass changes) and dissipation (surface rigidity) changes at the nanoscale on the lipid bilayer surface. A Q-Sense E4 system (Biolin Scientific, Sweden) was used to perform these measurements. The QCM-D contains four flow modules in which the quartz crystals are placed between two gold electrodes. Solutions are then delivered at a constant flow rate (0.15 mL/min) through a peristaltic pump. The frequency and mass changes are measured at five different harmonics, called overtones (3, 5, 7, 9 and 11). The overtones allow for an understanding across the depth of the bilayer, where the third overtone is closest to the bilayer surface and the eleventh overtone is closest to the silica substrate. Each overtone of the sensor crystal's natural frequency (5 MHz) were measured and automatically normalized for each overtone (f/n) by the Q-Sense software.

Quartz crystals could be used up to ten times under favorable conditions. QCM-D sensor crystals (5 Hz), reactively sputter-coated with silicon dioxide, were purchased from Biolin Scientific (Gothenburg, Sweden). Crystals are cleaned with rinses including an ethanol, de-ionized water, and 2 % sodium dodecyl sulfate rinses. Following this, another rinse of de-ionized water is flowed over the crystal surface, followed by an air flow. Crystals are thoroughly dried with nitrogen gas before undergoing two cycles of 45 s oxygen plasma cleaning using Plasma Prep II (SPI Supplies, West Chester, PA). The oxygen plasma cleaning ensures the removal of any organic contaminants and slightly oxidizes the sensor surface.

Once the crystals are thoroughly cleaned, the experimental procedure begins with setting a baseline in Tris NaCl buffer. In order to form a stable lipid bilayer on a QCM-D sensor crystal, PC vesicle solution will be injected into the QCM-D chamber at 0.15 mL/min for 8 min to allow vesicles to attach to the silica-coated crystal surface. In order to remove un-ruptured vesicles, the SLB was rinsed with buffer for 6 min. The process of forming an SLB using QCM-D has been previously established [15–19]. In preparation for the gold NP solution, which is in a bulk media of de-ionized water, the crystal surface is first rinsed with water for 8 min in order to establish a similar viscosity. After that, the gold NP solution is allowed to interact with the bilayer at the three designated concentrations for 10 min. At least five replicates were performed for each concentration of the 2 nm gold NPs.

Once frequency changes were measured, the Sauerbrey relationship can be applied in order to obtain mass change values:

$$\Delta \mathrm{m} = -\frac{\mathrm{C} \cdot \Delta \mathrm{f}}{\mathrm{n}}$$

Where C is the mass sensitivity constant C = 17.7 ng cm-2 Hz-1. For the analysis, Δf data for overtones 3, 5, 7, 9, and 11 were measured. The Sauerbrey model is used for rigid systems and to ensure the validity of using this model the system must experience a low dissipation value. Also, the mass adsorbed must be small relative to the quartz crystal and evenly distributed over the crystal area [20]. The Sauerbrey relationship may underestimate the mass for soft films because they do not couple completely to the sensor crystal, but it calculates a close approximation of mass change of a rigid surface [21, 22].

5.3 Results and Discussion

5.3.1 Bilayer Formation and Gold NP Interaction

QCM-D was used to monitor vesicle attachment and rupture on the silica crystal surface. Frequency (blue) and dissipation (red) values were measured by the QCM-D on the nanoscale in real time during each step of the process (Fig. 5.1). A representative plot at a gold NP concentration of 1.0×10^{12} NPs/mL is shown in Fig. 5.1. A frequency decrease (mass addition) was first observed during vesicle adsorption to the surface (a). Rapidly following this adsorption, the vesicles ruptured which formed the bilayer and caused a frequency increase (mass decrease). During this process, dissipation first increased due to the formation of a soft film followed by a decrease due to rigid film formation when the vesicles ruptured and released fluid. Next, in step (b), buffer was rinsed over the bilayer to remove any un-ruptured vesicles and stabilize the system. Since the gold NP solvent solution was water, water was first rinsed through the QCM-D in order to create a baseline prior to the addition of NPs to the system (c). 2 nm gold NPs then interacted with the SLB for 10 min at three different concentrations (d). Water and buffer were then rinsed again, respectively, to complete the experiment (e and f).

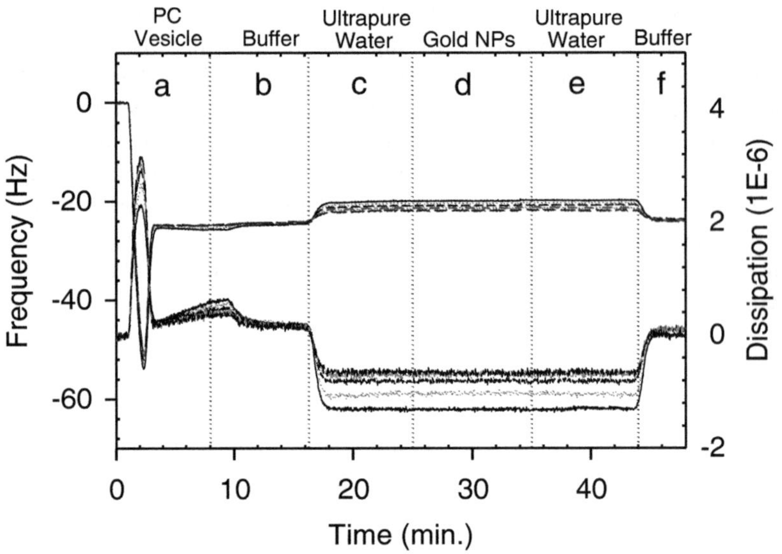

Fig. 5.1 Representative plot for QCM-D frequency (*blue lines*) and dissipation (*red lines*) changes versus time. Overtones 3, 5, 7, 9, and 11 are shown. (**a**) Bilayer formation occurs during flow of PC for the first 8 min. of the experiment. (**b**) Tris NaCl buffer then stabilized the bilayer and removed any un-ruptured vesicles. (**c**) In order to account for the bulk solution the NPs are suspended in, water was then flowed through the system for 10 min prior to their exposure. (**d**) The representative plot shows results for 2 nm gold NPs at a concentration of 1.0×10^{12} particles/mL interacting with the bilayer. Experiments were also performed at concentrations of 5.0×10^{10} and 1.0×10^{14}. (**e**) Water and (**f**) buffer were then flowed again in order to complete the experiment. The time for each step was buffer rinse 8:09; water rinse 14:34; gold NPs 23:06; water rinse 33:44; buffer 42:08

5.3.2 *Effect of Varying Concentration on NP-membrane Interactions*

Using the Sauerbrey relationship, mass changes were calculated from the frequency changes during the 10 min NP interaction with the lipid bilayer (Fig. 5.2). Three concentrations of 5×10^{10}, 1×10^{12}, and 1×10^{14} particles/mL were examined and the trend for mass addition and removal changed at higher concentrations. NPs suspended at the lowest concentration exhibited the most lipid perturbation. The middle concentration (1×10^{12} particles/mL) also showed some lipid perturbation, similar to the lowest concentration but not in the same magnitude. The results for the highest concentration yielded an opposite mechanism, with small amounts of mass added to the lipid surface at this concentration.

According to the changes in dissipation for all of the three concentrations, the dissipation changes are very low, showing that the bilayer is rigid, which aids in supporting the use of the Sauerbrey model. Also, small dissipation changes during the period of NP interaction with the bilayer indicate that the bilayer remained rigid during the experiments (Fig. 5.3).

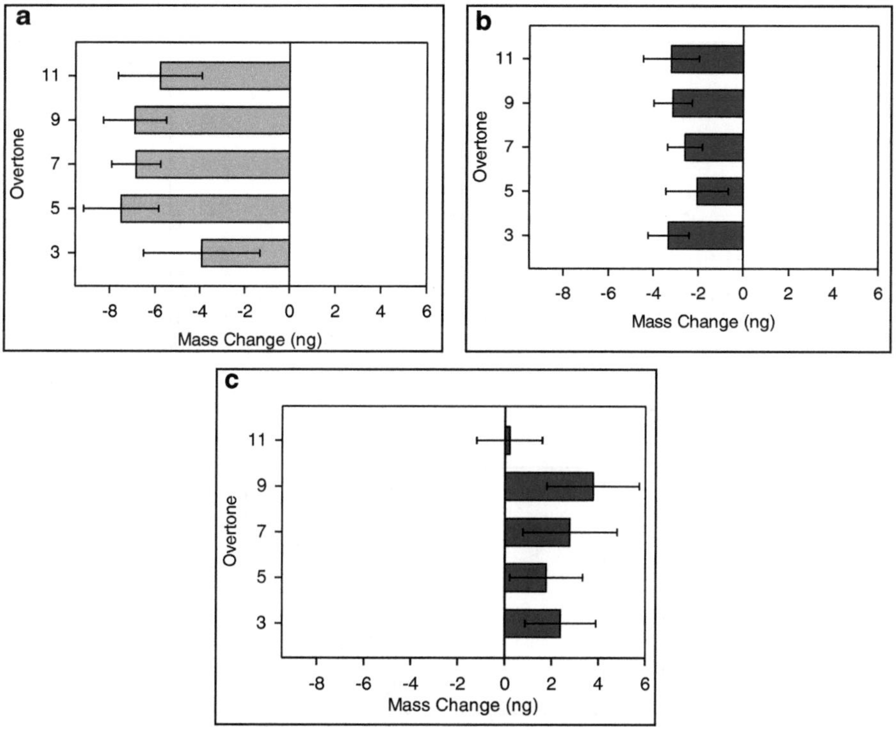

Fig. 5.2 Mass changes for nanoparticle-membrane interaction with the applied Sauerbrey relationship for 2 nm gold NPs at concentrations of (**a**) 5.0×10^{10}; (**b**) 1.0×10^{12}; and (**c**) 1.0×10^{14} particles/mL. Particles are suspended in a bulk solution of water at a neutral pH of 7. At least five experimental samples were performed for each concentration

The relevance of studying nanoparticle concentration is based on the growing disposal rates, leading to larger and varying concentration values in the environment [23, 24]. The consequences depending on the concentration of particles is unknown, and more specifically it is further unknown depending on size and material [25]. Our results have indicated that a two order magnitude change in NP concentration can alter the interaction of particles with a lipid bilayer.

The effect of concentration was determined for 2 nm gold NPs on the interaction with a supported lipid bilayer. Our data presents a mechanistic change in between the concentration values of 1.0×10^{12} particles/mL and 1.0×10^{14} particles/mL where the mechanism changes from mass loss to mass addition, respectively. While the magnitude of these mass changes are small, this shift may be important in understanding the threshold for NP perturbation and accumulation with a supported lipid bilayer.

Effects of size, concentration and exposure time on gold nanoparticles cellular toxicity and recovery were investigated in a study conducted by Mironva et al. [8]. The interactions of gold nanoparticles with human dermal fibroblasts are described

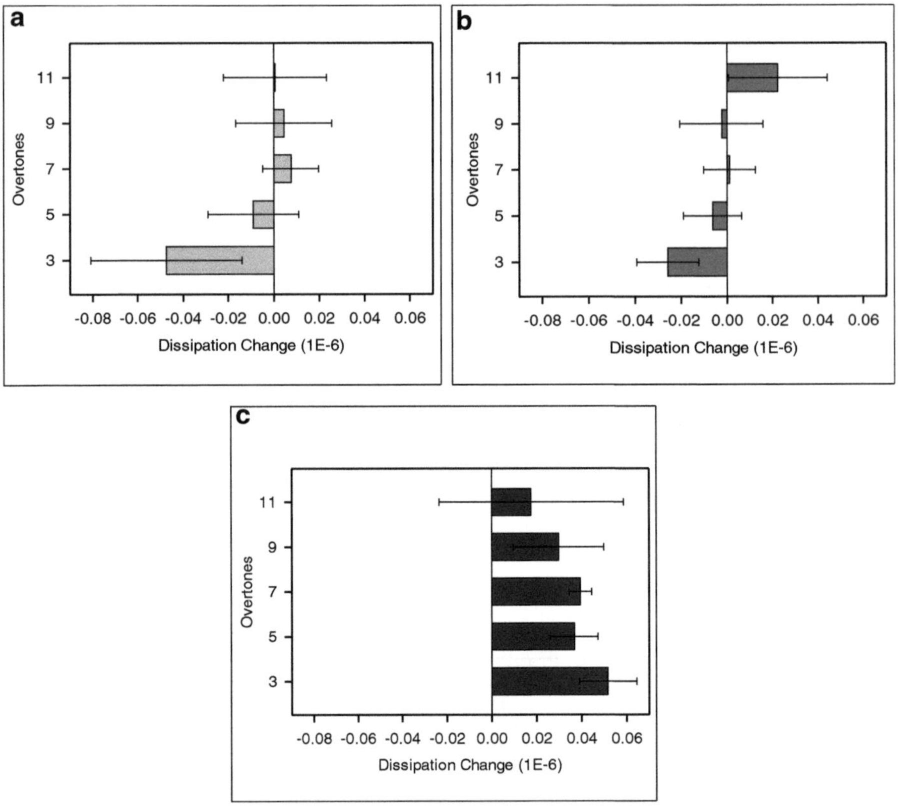

Fig. 5.3 Dissipation changes for 2 nm gold NPs at concentrations of (**a**) 5.0×10^{10}; (**b**) 1.0×10^{12}; and (**c**) 1.0×10^{14} particles/mL. At least five experimental samples were performed for each concentration

and it is shown that NPs can penetrate the plasma membrane and accumulate in large vacuoles. It is also demonstrated that the uptake of the gold nanoparticles is a function of time, their size and concentration. A decrease in growth rate was observed with increasing gold nanoparticles concentration. This effect was also observed in another study done by Goodman et al. [10]. In this study, they have done dye release studies using lipid vesicles and observed concentration-dependent lysis as the lysis of the negative SOPC/SOPS vesicles increased with increasing concentration of mixed monolayer protected gold clusters.

In addition to NPs concentration, other parameters such as NPs size and coating have been widely studied. Smaller sized particles are often considered as more dangerous as the smaller the particles, the higher the surface area to volume ratio which leads to more interaction of the NPs with cellular or intracellular components [26]. To have better mechanistic understanding of effect of NPs size on NPs-cell interactions, we have previously used QCM-D technique in our lab. NPs with different diameters of 2, 5, 10, and 40 nm were studied. The results for all four sizes of citric acid-stabilized gold NPs in water showed positive frequency changes, corresponding to net mass loss from lipid removal.

In most of the natural systems, in addition to nanoparticles and cells, there are other materials affecting the nanoparticle-cell interactions. For example in a biological fluid, NPs rapidly get covered by biomolecules, such as proteins and lipids, forming a biomolecular corona that effectively screens the bare NP surface. Therefore it is interesting to consider more realistic circumstances and consider not the bare NP surface that interacts with the cell but the NP-biomolecule complex. For instance in an study conducted by Lesniak et al. it was shown that the adhesion properties of NPs to the cell membrane is strongly affected by the presence or absence of serum and this accounts, at least partially, for a corresponding difference in NP-uptake levels [27]. The same situation is valid for NPs in the environment. In the environment, there are different natural organic matters which affect the interactions between nanoparticles and cells. Further in our research, in order to understand NP interaction in the environment, the application of humic acid materials was added. Humic acids are the major components of natural organic matter in soil and water. We have previously looked at the effect of gold NPs in the presence of Aldrich humic acid, Suwanee River humic acid, Fulvic acid, and Elliot Soil humic acid.

Acknowledgements This work was supported in part by the National Science Foundation (CBET 0966496). For help with concentration experiments, the authors thank Jeniece Macedonio for her contribution.

References

1. Barreto JA, O'Malley W, Kubeil M, Graham B, Stephan H, Spiccia L (2011) Nanomaterials: applications in cancer imaging and therapy. Adv Mater (Deerfield Beach, Fla) 23(12):H18–H40. doi:10.1002/adma.201100140
2. Wang S, Lu W, Tovmachenko O, Rai US, Yu H, Ray PC (2008) Challenge in understanding size and shape dependent toxicity of gold nanomaterials in human skin keratinocytes. Chem Phys Lett 463(1–3):145–149. doi:10.1016/j.cplett.2008.08.039
3. Yen H-J, Hsu S-H, Tsai C-L (2009) Cytotoxicity and immunological response of gold and silver nanoparticles of different sizes. Small (Weinheim an Der Bergstrasse, Germany) 5(13):1553–1561. doi:10.1002/smll.200900126
4. Nel AE, Mädler L, Velegol D, Xia T, Hoek EMV, Somasundaran P, Thompson M (2009) Understanding biophysicochemical interactions at the nano-bio interface. Nat Mater 8(7):543–557. doi:10.1038/nmat2442
5. Tarantola M, Pietuch A, Schneider D, Rother J, Sunnick E, Rosman C, Janshoff A (2011) Toxicity of gold-nanoparticles: synergistic effects of shape and surface functionalization on micromotility of epithelial cells. Nanotoxicology 5(2):254–268. doi:10.3109/17435390.2010.528847
6. Hou W-C, Moghadam BY, Corredor C, Westerhoff P, Posner JD (2012) Distribution of functionalized gold nanoparticles between water and lipid bilayers as model cell membranes. Environ Sci Technol 46(3):1869–1876. doi:10.1021/es203661k
7. Pan Y, Neuss S, Leifert A, Fischler M, Wen F, Simon U, Jahnen-Dechent W (2007) Size-dependent cytotoxicity of gold nanoparticles. Small (Weinheim an Der Bergstrasse, Germany) 3(11):1941–1949. doi:10.1002/smll.200700378

8. Mironava T, Hadjiargyrou M, Simon M, Jurukovski V, Rafailovich MH (2010) Gold nanoparticles cellular toxicity and recovery: effect of size, concentration and exposure time. Nanotoxicology 4(1):120–137. doi:10.3109/17435390903471463
9. Zhang Y, Chen Y, Westerhoff P, Crittenden J (2009) Impact of natural organic matter and divalent cations on the stability of aqueous nanoparticles. Water Res 43(17):4249–4257. doi:10.1016/j.watres.2009.06.005
10. Goodman CM, McCusker CD, Yilmaz T, Rotello VM (2004) Toxicity of gold nanoparticles functionalized with cationic and anionic side chains. Bioconjug Chem 15(4):897–900. doi:10.1021/bc049951i
11. Stankus DP, Lohse SE, Hutchison JE, Nason J (2011) Interactions between natural organic matter and gold nanoparticles stabilized with different organic capping agents. Environ Sci Technol 45(8):3238–3244. doi:10.1021/es102603p
12. Sutton R, Sposito G (2005) Molecular structure in soil humic substances: the new view. Crit Rev 510:9009–9015
13. Hardy GJ, Nayak R, Zauscher S (2013) Model cell membranes: techniques to form complex biomimetic supported lipid bilayers via vesicle fusion. Curr Opin Colloid Interface Sci 18(5):448–458. doi:10.1016/j.cocis.2013.06.004
14. McCubbin GA, Praporski S, Piantavigna S, Knappe D, Hoffmann R, Bowie JH, Martin LL (2011) QCM-D fingerprinting of membrane-active peptides. Eur Biophys J□: EBJ 40(4):437–446. doi:10.1007/s00249-010-0652-5
15. Barenholz Y, Gibbes D, Litman BJ, Goll J, Thompson TE, Carlson RD (1977) A simple method for the preparation of homogeneous phospholipid vesicles. Biochemistry 16(12):2806–2810, Retrieved from http://www.ncbi.nlm.nih.gov/pubmed/889789
16. Keller CA, Glasmästar K, Zhdanov VP, Kasemo B (2000) Formation of supported membranes from vesicles. Phys Rev Lett 84(23):5443–5446, Retrieved from http://www.ncbi.nlm.nih.gov/pubmed/10990964
17. Cremer PS, Boxer SG (1999) Formation and spreading of lipid bilayers on planar glass supports. J Phys Chem 291:2554–2559
18. Mechler A, Praporski S, Atmuri K, Boland M, Separovic F, Martin LL (2007) Specific and selective peptide-membrane interactions revealed using quartz crystal microbalance. Biophys J 93(11):3907–3916. doi:10.1529/biophysj.107.116525
19. Richter RP, Brisson AR (2005) Following the formation of supported lipid bilayers on mica: a study combining AFM, QCM-D, and ellipsometry. Biophys J 88(5):3422–3433. doi:10.1529/biophysj.104.053728
20. Dixon MC (2008) Quartz crystal microbalance with dissipation monitoring: enabling real-time characterization of biological materials and their interactions. J Biomol Tech 19:151–158
21. Furman O, Usenko S, Lau BLT (2013) Relative importance of the humic and fulvic fractions of natural organic matter in the aggregation and deposition of silver nanoparticles. Environ Sci Technol 47(3):1349–1356. doi:10.1021/es303275g
22. Vogt BD, Lin EK, Wu W, White CC (2004) Effect of film thickness on the validity of the Sauerbrey equation for hydrated polyelectrolyte films. J Phys Chem B 108:12685–12690
23. Gottschalk F, Nowack B (2011) The release of engineered nanomaterials to the environment. J Environ Monit: JEM 13(5):1145–1155. doi:10.1039/c0em00547a
24. Lewinski N, Colvin V, Drezek R (2008) Cytotoxicity of nanoparticles. Small (Weinheim an Der Bergstrasse, Germany) 4(1):26–49. doi:10.1002/smll.200700595
25. Thomas CR, George S, Horst AM, Ji Z, Miller RJ, Peralta-Videa JR, Zink JI (2011) Nanomaterials in the environment: from materials to high-throughput screening to organisms. ACS Nano 5(1):13–20. doi:10.1021/nn1034857
26. Soenen SJ, Manshian B, Montenegro JM, Amin F, Meermann B, Thiron T, Cornelissen M, Vanhaecke F, Doak S, Parak WJ, De Smedt S, Braeckmans K (2012) Cytotoxic effects of gold nanoparticles: A multiparametric study. ACS Nano 7:5767–5783
27. Lesniak A, Salvati A, Santos-Martinez MJ, Radomski MW, Dawson K, Åberg C (2013) Nanoparticle adhesion to the cell membrane and its effect on nanoparticle uptake efficiency. J Am Chem Soc 135(4):1438–1444. doi:10.1021/ja309812z

Chapter 6
Application of Cost-Effective Biological Tools for Assessing of Chemical Poisoning

Olga Malev, Roberta Sauerborn Klobučar, Tatjana Tišler, Damjana Drobne, and Polonca Trebše

Abstract There is an enormous amount of new chemicals emerging every day with potential adverse effects for humans and environment. To assess these chemicals we need cost effective and reliable biological tools. Here we present in detail one of many bioassays suited for assessing chemical poisoning in the environment. This is a terrestrial isopod single-species test. Advantages of this test with terrestrial isopods are its flexibility in terms of exposure duration and versatile biomarker selection. By alternating test duration, one can test also those substances which are with moderate or low toxic potential; while a variety of biomarkers at different levels of biological complexity increase the relevance of tests results. In addition to laboratory single-species tests, terrestrial isopods could also be used for biomonitoring of pollutants. So far, this was thoroughly assessed in case of metal pollution, although isopods have a potential to be used also for biomonitoring of some organic substances and metal based nanoparticles. This is usually not the case in other

---

O. Malev
Laboratory for Environmental Research, University of Nova Gorica, Vipavska 13, PO Box 301, SI-5000 Nova Gorica, Slovenia

R.S. Klobučar
Division of Materials Chemistry, Laboratory of Ichtiopatology – Biological Materials, Ruđer Bošković Institute, Bijenička 54, 10001 Zagreb, Croatia

T. Tišler
National Institute of Chemistry, Laboratory for Environmental Sciences and Engineering, Hajdrihova 19, SI-1000 Ljubljana, Slovenia

D. Drobne
Biotechnical Faculty, Department of Biology, University of Ljubljana, Večna pot 111, SI-1000 Ljubljana, Slovenia

P. Trebše (✉)
Laboratory for Environmental Research, University of Nova Gorica, Vipavska 13, PO Box 301, SI–5000 Nova Gorica, Slovenia

Faculty of Health Sciences, University of Ljubljana, Zdravstvena pot 5, SI-1000 Ljubljana, Slovenia
e-mail: polonca.trebse@ung.si; polonca.trebse@zf.uni-lj.si

T.A. Camesano (ed.), *Nanotechnology to Aid Chemical and Biological Defense*, NATO Science for Peace and Security Series A: Chemistry and Biology,
DOI 10.1007/978-94-017-7218-1_6

standardised ecotoxicity tests species. In addition, we discuss that at present, we do not need completely new tests for emerging substances like new generations pesticides, personal care products and products of nanotechnologies, but is sufficient to adopt and modify existing test protocols.

6.1 Introduction

Chemical, biological, radiological and nuclear defense (CBRN defense or CBRND) consists of passive protection, contamination avoidance and CBRN mitigation. To avoid any of these hazards the methods should be available to detect chemicals presence and their hazardous potential.

In the environment, bioassays and chemical analysis yield complementary information for hazard characterisation as well as for risk assessment. Bioassays indicate the effect of pollutants, while chemical analyses aim to quantify components of the mixture, but can never identify the entire cocktail of pollutants. These cocktails can include pesticides, biocides, pharmaceuticals, personal care and consumer products while recently a lot of attention is given to product of nanotechnologies [1–3].

6.2 Hazard Characterisation of New Substances

After the II World War, together with a huge development and production of new chemical substances, the growing concern about their environmental effects has raised. Toxic effects of different xenobiotics in the environment depend on their chemistry and their fate in the environment. Among anthropogenic chemicals daily released into the environment, special attention should be given to persistent chemicals especially those with bioaccumulation potential and/or with a diverse range of transformation products (TPs), which can be even more toxic [4] and more persistent than the parent compound [5].

Attention should be given also to new compounds with unknown biological effects. Among industrial chemicals (chlorinated solvents, petrochemicals, polychlorinated biphenyls, organotin compounds, asbestos), pesticides belong to a broad and diverse group of compounds. Their extensive use to enlarge the production in agriculture has led to growing input of these pollutants into environment over the last decades. Organophosphorous pesticides (OP) have been among most extensively used pesticides, with the restriction on the use of organochlorine pesticides (OC). OPs and carbamate pesticides are generally more acutely toxic and exert potentially more widespread effects on non-target organisms than organochlorine pesticides do. Since pests relatively quickly develop resistance to existing insecticides, the discovery of new insecticidal compounds with unique modes of action is essential for efficient crop protection in the future. Neonicotinoids are relatively new systemic insecticides which are chemically similar to nicotine – the toxin present in tobacco.

With partially positive charge they can irreversibly bind to nAChRs which is higher for nAChRs from insects than those from mammals [6, 7]. In the last years new groups of insecticides are coming to the market, for example anthranilic diamides, representing a class of insecticides with a novel mode of action, ryanodine receptor activation [8].

Cosmetics (personal care products e.g. sunscreen, lipsticks, shampoos and hair sprays) are a part of a broad and diverse group of chemicals which are potentially polluting the environment. For example, cosmetics containing UV filters tremendously expand on the market as a result of the growing awareness of the harmful exposure to the sun and the consequent increased risk of morbidity for skin cancer. In addition to inorganic pigments, which reflect UV light in particular, organic compounds that absorb UV light are also used. Because UV light is of a fairly broad spectral range, 400–290 nm (UVA and UVB), no compound is capable of preventing the exposure to the whole spectrum by itself. Therefore, a combination of several compounds that cover the whole area is usually applied. UV filters must be relatively stable when exposed to UV radiation, however, several recent studies showed that these compounds are decomposed by light by direct photolytic reactions, or by chlorination of aromatic rings or side chain, which is due to the presence of chlorine and chlorate medium (such as pools, salty sea water). Information on the fate of these compounds in the environment is scarce for most of these chemicals, especially for chlorinated products.

Most recently, the nanotechnologies design and manufacture materials on the scale of one-billionth of a meter with completely new desirable characteristics. The variety of nanomaterials is great, and their range of properties and possible applications appear to be enormous, from extraordinarily tiny electronic devices, including miniature batteries, to biomedical uses, and as packaging films, superabsorbents, components of armour, and parts of automobiles as well as in cosmetics, food and cloth industry, electronics, biomaterials and energy production. Such wide application will eventually lead to substantial environmental burden what will pose a potential risk to biota due to their high reactivity.

6.3 Aim of the Paper

In our paper, we present an example of a toxicity tests with terrestrial isopods, which is highly adoptive in terms of duration and biomarker selection, it is cost effective, could be performed in the laboratory or in the field. We discuss whether new tests are needed or whether it is better when existing test protocols are adopted for emerging substances which include new generations of pesticides, cosmetics and products of nanotechnologies.

6.4 Exposure Assessment and Hazard Characterisation

Exposure to pollutants is generally defined as the contact between chemicals and a receptor and can be characterised either by direct (organism monitoring or biological markers) or indirect methods (environmental monitoring or modelling). The direct measures of exposure can be considered as the best, however also indirect methods have rapidly evolved in the last years, especially due to the available computer models which can simulate environmental pollution.

Exposure assessment (e.g. exposure concentrations and related dose for specific pathways) is, together with effects assessment (i.e. hazard identification, dose-response evaluation), an integral part of the risk assessment process. It is composed of non-biological analyses of behaviour of pollutants (e.g. engineered nanoparticles (ENPs)) under environmental or test conditions and of biological methods which elucidate their bioavailability or their dissolution products, bioaccumulation potential and persistence once they enter the body. The major goal of exposure assessment and hazard characterisation is to provide a qualitative and quantitative description of the agent's contact with humans or environmental organisms (Fig. 6.1). Different methods are developed for this purpose and their selection depends on the substance of interest. Much of the variation in risk estimates is associated with variation in exposure classification. For different substances, before tests protocols are widely applied, reference compounds and tests methods are selected or elaborated. In case of nanoparticles as a very new compound with toxicological potential, reference exposure assessment methods are under development within NanoValid EU large-scale project [9].

6.5 Conventional Bioassays for Toxicity Evaluation

Both acute and chronic tests are relatively expensive as they require the continuous culturing and maintenance of live test organisms, restricting their application to specialised laboratories and operators [10]. This has led to the development of a range of rapid microbiotests. In addition to the microbiotests, modified "traditional" tests have also been evaluated by a number of researchers. Among most important modifications of existing conventional tests is the selection of measured endpoints. When the mode of action is unknown or when a mixture of substances is tested, a variety of endpoints is needed along with possibility to alter exposure duration. For example, when the tested substances are of minor significance the organism is able to continue its functions in a normal manner. However, under conditions of additional stress, such as changes in pH, dissolved oxygen or temperature, the effects induced by a tested substance can be detected because the organism do not retain its normal activity. Effects of less biologically potent chemicals could be detected when more sensitive endpoints are selected or when exposure is prolonged.

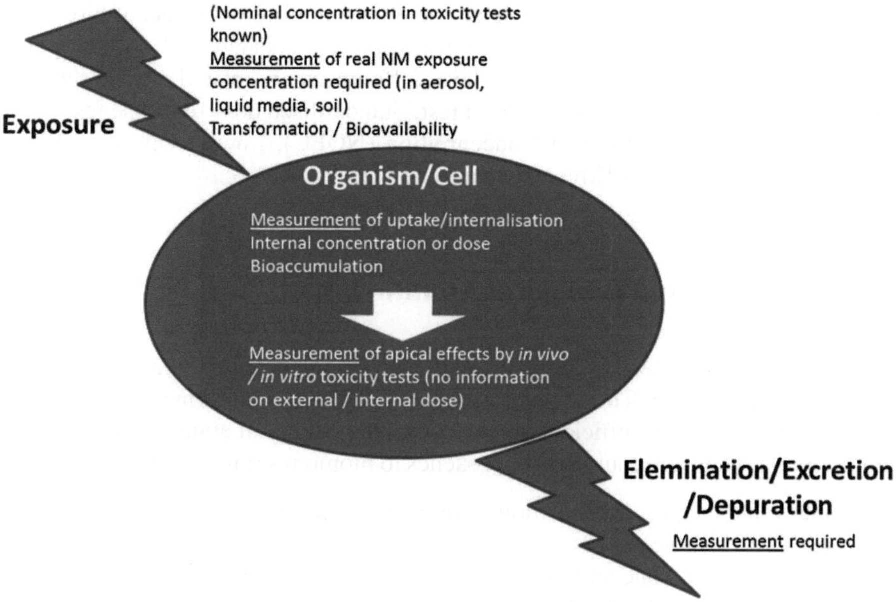

Fig. 6.1 Overview – external and internal exposure, as well as elimination have to be considered. Exposure assessment can be done by quantitative as well as qualitative methods; however, quantitative information will be more useful to establish a link to hazard (The figure is adopted from NANOVALID report prepared by D. Drobne)

With each new substance a question appears if existing tests are suited also for new products and it is also questioned their sensitivity and biomarker selection.

A lot of studies have been made regarding water quality criteria and hazard assessment of chemicals in the aquatic environment, the same is not true for the terrestrial ecosystem. Consequently much less ecotoxicity tests and data sets exist for terrestrial environment. In general, ecotoxicity tests are used for different purposes like environmental risk assessment (ERA) or chemical regulation. To discuss the criteria for being a good test, the purpose of testing should be well defined. The ERA includes several distinct steps with different purposes. In the first step, adverse effects, which a substance has an inherent capacity to cause at certain concentration, are identified. This step is called hazard identification step. Afterwards the relationship between dose of exposure to a substance and an effect are estimated in a dose response assessment with standardised base set of ecotoxicity tests and potential environmental exposure levels are analysed (exposure assessment). Finally, risk characterisation is performed by estimation of the incidence and severity of the adverse effects likely to occur in environment (i.e. the quantification of that likelihood). For dose response assessment, a standardised set of toxicity tests is required, but hazard identification, a step whose purpose is to provide a holistic view of the potential hazard of the substance in question, is more flexible in terms of using different tools. In hazard identification, it is crucial to have as much information as possible on the effects at different levels of biological organisation. Tests used for hazard identification and standardised dose response tests have to fulfil different criteria. These differences are related also to the type of toxicity data which are provided by a test. Standardised dose response tests have to provide No Observed Effect Concentration (NOEC), Lowest Observed Effect Concentration (LOEC) or Effect Concentration (EC_x) data.

6.6 Chemical and Biological Monitoring

In addition to toxicity testing, the information on the impact of pollutants is obtained also by biomonitoring. The objective of environmental monitoring (chemical and biological) is to collect sufficient data to assess the quality of aquatic and terrestrial environment. There are four main approaches to biomonitoring, which are based on:

- changes in species composition, presence or absence of species from the investigated sample site;
- measurements of concentrations of chemicals in organisms;
- assessment of the effects of chemicals on organisms and
- detection of genetically different strains of species, which have evolved resistance in response to chemicals.

In the European Union, biomonitoring based on measurements of levels of priority substances which tend to bioaccumulate in biota (EC/2008/105) and quantification of effects of chemicals on organisms by using a wide range of different biomarkers [11, 12] and corresponding protocols is involved in monitoring programme of environmental compartments (such as surface waters and seawater quality).

6.7 Problem Description

A variety of short-term and long-term toxicity test methods are designed to test effects of different substances and extrapolate observed effects on humans and environment. In 1951s the need to develop uniform, standardised test methods was recognised to improve the comparability of test data. A standardised test method, which includes reference toxicants, standard test species and detailed test protocol (standard operational procedures -SOPs), help to improve comparability, replicability and reliability of toxicity test results [13]. But on the other side, standardisation prevents modifications of tests protocols in order to increase the relevance of the outcome. Such modifications are of high importance when substances with unknown mode of action are tested or when effects of mixtures are assessed.

6.8 Terrestrial Isopod Bioassay for Conventional and New Substances

Here we present in detail a terrestrial isopod toxicity test which was successfully used for toxicity testing as well as biomonitoring of bioavailable pollutants [14] and has a potential to be modified also for testing effects and bioaccumulation of new substances. The extensive research of recent years in terrestrial ecotoxicology identified earthworms, mites, isopods and collembolans as most relevant for dose response tests [15, 16]. Ecotoxicity test with terrestrial isopods (*Isopoda*, Crustacea) are applicable for chemical and nanoparticles hazard profiling (hazard identification) as well as single species dose response tests. In the following text the applicability and characteristics of a dose response test with terrestrial isopods are presented in some detail.

6.9 Terrestrial Isopods in Ecotoxicity Studies

Terrestrial isopods are widely accepted as organisms of choice in terrestrial ecotoxicology and eco-physiology [15, 17, 18]. They are used also for assessing sustainability of the terrestrial environment [19]. There are up to hundred scientific papers reporting the results of studies on pollutants or single chemicals and particles on isopods (ISI Web of Knowledge). Among most popular species in these kind of research is a terrestrial isopods *Porcellio scaber*. Many ecotoxicity studies are conducted also with *Porcellio dilatatus*, *Oniscus asellus* or *Porcellionides pruinosus*. These studies report the ecotoxicity of variety of chemical pollutants on terrestrial isopods. Animals are usually exposed to food or substratum dosed with different concentrations of tested chemicals for some weeks and responses at different levels of biological complexity ranging from enzyme induction/inhibition, histological changes, and physiological level response up to growth, reproduction and mortality are recorded. Among reasons for popularity of isopods in such studies is the fact that they are relatively large to conduct analyses on individual animals or even tissue, they are abundant, relatively simple to identify, and simple aging. In addition, there is sufficient background knowledge on the biology and ecology of this species. Still, terrestrial isopods are usually not listed as an ecotoxicity test of choice for dose response assessment. This discrepancy might arise from the fact that isopod tests are not standardised and that each laboratory conducts the research under different conditions and therefore despite high quality toxicity data they could not be used in ERA. For example, LOEC or EC_x values could not be compared if the duration of a test is different and if exposure scenario or food preparation is not the same. Obviously the scientist become aware of the need to provide data which should be useful for further processing for ERA, therefore in some papers the same laboratory conditions, food preparation and duration of expose was used. In such a caste data could be compared, combined, and further used. In the following text these studies are presented in more details.

6.10 Single Species Dose Response Toxicity Tests with Terrestrial Isopods

Isopods also allow measuring biomarkers at different levels of biological complexity. Biomarkers may be divided into biomarkers of exposure and biomarkers of effect. By a definition posed by Kammenga et al. [20], biomarkers of effect are those measured alterations in an organism that are associated with possible health impairment or disease. Biomarkers of exposure are those measured responses of an organism that indicate the interaction between a xenobiotic agent and some target molecule or cell [21].

6.10.1 Biochemical Biomarkers

Most frequently analysed biomarkers in toxicity studies with isopods are changes in different enzyme activities; changes in energy reserves, total protein content [22, 23]. Biochemical biomarkers are expected to be more sensitive than changes observed at the whole-organism level. To evaluate the sensitivity of biochemical biomarkers, one should compare the LOEC values of biochemical biomarkers to the LOEC values obtained for physiological biomarkers, such as feeding, growth change and survival of isopods. Toxicity data for isopods are combined in Table 6.1.

From the comparison of toxicity data one can conclude that biochemical biomarkers are not always more sensitive as higher level end-points. Their sensitivity depends on the (a) mode of action of a chemical, (b) its duration and (c) it is species-specific. It has previously also been suggested that the high variability in the responsiveness of biochemical biomarkers is a function of the species being investigated, the kind of exposure, the class of chemicals, and the duration of exposure. This sensitivity of biochemical biomarkers has important consequences for their use in the evaluation of risk or hazard. The use of a single biochemical biomarker in risk calculation would over- or under-estimate the risk. The relevant risk estimation can only be calculated when a battery of biomarkers are assessed. Further, biochemical biomarkers can be used for ranging of chemicals according to their potential to cause adverse effects and to characterise their hazard. Comparative toxicity data are of high importance in order to direct further toxicity investigations.

6.10.2 Mortality, Growth, and Reproduction Changes

The most widely used toxicological endpoints are mortality, growth, and reproduction. The problem with mortality as a toxicological endpoint is its low sensitivity. The use of other responses such as growth and reproduction is sometimes not convenient because of the long duration of the required test [17].

Table 6.1 The toxicity data obtained for adult isopods [22, 23]

	Short-term exposure (3 days)			Long-term exposure (14 days)	
	Cd^{2+} ($CdCl_2$)[a]	Zn^{2+} ($ZnCl_2$)[a]	Cu^{2+} ($CuCl_2$)[a]	Diazinon[a]	IMI[b]
	500	5000	5000	100	25
LOEC values (µg/L)					
Biochemical biomarkers	100	1000	>5000	50	25
Physiological biomarkers	>500	5000	5000	100	10
Combined (biochem. + physiol.)	100	1000	5000	50	10
NOEC values (µg/L)					
Biochemical biomarkers	10	<1000	5000	10	10
Physiological biomarkers	500	2000	2000	50	<10
Combined (biochem. + physiol.)	10	<1000	2000	10	<10
Toxicity ranking LOEC					
Biochemical biomarkers	$Cd^{2+} < Zn^{2+} < Cu^{2+}$			IMI < diazinon	
Physiological biomarkers	Cd^{2+}[c], $Zn^{2+} = Cu^{2+}$			IMI < diazinon	
Combined (biochem. + physiol.)	$Cd^{2+} < Zn^{2+} < Cu^{2+}$			IMI < diazinon	
Toxicity ranking NOEC					
Biochemical biomarkers	$Cd^{2+} < Zn^{2+} < Cu^{2+}$			IMI = diazinon	
Physiological biomarkers	$Cd^{2+} < Zn^{2+} = Cu^{2+}$			IMI < diazinon	
Combined (biochem. + physiol.)	$Cd^{2+} < Zn^{2+} < Cu^{2+}$			IMI = diazinon	

IMI imidacloprid, *LOEC* Lowest Observed Effect Concentrations, *NOEC* No Observed Effect Concentrations

[a]Stanek et al. [24]

[b]Drobne et al. [22]; > the LOEC was not observed up to the highest concentration tested

[c]Ranking could not be determined

6.10.3 Feeding Activity

One approach to assess food quality is by using feeding activity/inhibition of macroinvertebrates as a measure for a wide range of stressors, and as it can be measured ex situ and in situ, it has been proposed as an indicator for food quality [25]. As feeding rate is variable, depending on the status of the organism and its surrounding environment, understanding those factors and the causes for variability is crucial for sample/chemical toxicity assessment. Feeding experiments can be expanded by including food choice as an endpoint [26, 27]. Feeding assay is based on the consumption of leaf mass, an interesting parameter which could be used to assess potential changes in ecosystem processes. Feeding activity was also combined by other lower level parameters like biochemical biomarkers, (cholinesterase (ChE) and glutathione-*S*-transferase (GST)) or higher level biomarkers like weight change or mortality [23, 28]. Any changes in feeding rates of terrestrial isopods affect the decomposition process and subsequently matter and energy flux through ecosystems. Isopods have direct and indirect contribution to decomposition processes. First they comminute organic material and second they stimulate microbial activates. Therefore any changes in feeding activity of isopods would affect energy flow and soil turnover. The first ecotoxicological laboratory test for assessing the sub-lethal effects of chemicals on terrestrial isopods was based on changes of feeding activity as a toxicity endpoint [25, 29].

Subsequently the feeding response of juveniles or adults was investigated following exposure to many different chemicals and nanoparticles [22, 23, 30, 31]. The method employed is non-destructive, provides a rapid indication of the status of groups of individuals, and is based on a time-response analysis of the consumption of isopods and the determination of median feeding. The feeding activity of juvenile was found to be a sensitive response criterion for use in assessing the sub-lethal toxicity of metals and pesticides (see [23]). Reductions in isopod feeding activity were identified following 14 days exposure. A sustained reduction in feeding rates may cause growth inhibition and impaired reproduction. The advantage of measuring feeding activity as a response endpoint is that at the end of the test they could be re-exposed to the same toxic chemical again to study the effect of pulsed exposure or to study the recovery.

6.10.4 Physiological Biomarkers

In toxicity testing, a response of an organism to chemicals is measured by alteration in the function of one or more physiological processes or by the death of an organism. Death tells us relatively little about the mechanism of toxicity of chemicals. If toxicity is being evaluated, a sub-lethal response of the organism is desirable.

Response of the moult cycle to chemical toxicity was studied on *P. scaber* and proved to be very suitable response to metal contaminated food [32]. Moult cycle shows low variability in duration but illustrates a great range of responses to the toxicant. Moulting is sensitive even to low concentrations of zinc in the food, and its frequency is easily determined. The response of moulting frequency in single-species toxicity tests, chronic and acute, and in microcosm toxicity tests seems to be a promising measure of toxicity. Feeding response and moult frequency measure could be conducted in the same experiment and at the end the animals were not scarified.

6.10.5 Histological Biomarkers

The epithelium thickness is a reliable measure of the state of organisms, applied also on isopods [33–36]. In toxicity studies with isopods it was shown that epithelial thinning occurred after feeding with the Hg-dosed food. However, at the end of the 14-day feeding period, in the animals from the Hg polluted locations the epithelium was evidently thicker than at the beginning of the exposure. But, to induce histopathological changes as a result of feeding with chemically dosed food, most probably longer exposure time or higher doses are needed to provoke this or similar responses.

6.10.6 Cellular Biomarkers

Cells provide a link between molecular and biochemical events and physiological or whole-organism events. Promising methods that may offer a reliable way of assessing individual health status include assessments of lysosomal membrane stability (LMS). The derivation of lysosomal biomarkers was based on the assumption that because of their key role in cellular functioning, they should be indicative of the functional integrity of cells.

The lysosomal latency (LL) assay and the neutral red retention (NRR) assay are most frequently used for measuring LMS. This method is used on cell suspensions or cryostat sections of a tissue. In isopods the LMS was investigated on hepatopancreatic cells in animals exposed to pollutants in the laboratory or in the field populations [37, 38].

6.10.7 Linkage Biomarkers at Different Levels of Biological Complexity

One of the suggested criteria for the use of biochemical biomarkers in environmental pollution studies is that they should exhibit a link to adverse effects at organism level like growth and reproduction changes and mortality. Results of many authors confirm that certain changes of biochemical biomarkers are preceded by the effects at higher level of biological organization, but this is not a rule. Only in some rare studies linkage of a target enzyme inhibition to integrated organism response variables was demonstrated where animals were exposed through food [24, 39–41] or via substrate [42, 43]. It appears from a bulk of published data that a thorough toxicity study along levels of biological complexity in both juveniles and adults, in laboratory or field study would be a promising approach to provide comprehensive toxicity information. An integrated biomarker study conducted in parallel with different life stages of terrestrial isopods has not been conducted yet.

Physiological and histological endpoints are more sensitive to the same metal than whole-organism endpoints. Similar data indicating that whole-organism responses such as survival are less sensitive to elevated concentrations of chemicals in the food than biochemical or histological responses such as a 70 kD heat shock protein or ultra-structural changes of gland cells. Therefore, these endpoints could be used for early warning studies. Whole-organism responses are much more relevant in terms of environmental risk assessment.

6.11 Reference Toxicants and Reference Toxicological Endpoints

When terrestrial isopods are used as test organisms, positive controls are important because the animals can become tolerant to polluted environments. Tolerant populations will respond differently as non-tolerant. Therefore, it is necessary to test each new population of test organisms for sensitivity to pollutants before conducting toxicity testing. Although a laboratory culture must be maintained for (standard) toxicity tests, the sensitivity of this population to pollutants can also change, so it should be regularly tested with reference toxicants.

The results show that zinc chloride or zinc nitrate could be used as a reference toxicant in toxicity tests with *P. scaber* [17]. Zinc-positive controls could provide data on the sensitivity of different testing populations of terrestrial isopods. Studies with artificially contaminated substrates could be conducted with $ZnCl_2$, $Zn(NO_3)_2$, and ZnO.

Besides reference toxicants, reference endpoints could also be used as positive controls. The results with terrestrial isopods show that feeding rate is a very suitable parameter to be standardised as a toxicological endpoint in a positive control.

6.12 Data Acquisition and Processing

The outstanding advantage of performing toxicity tests with terrestrial isopods is that one can calculate effect and no effect levels of chemicals, because it is possible to determine the ingested amount of chemicals. This is the usual way of acquiring toxicity data with laboratory mammals but not the case with invertebrates.

Here an example is given how the author calculated chronic reference doses (RfDs) and hazard quotients (HQ). They calculated RfDs from No Observed Effect Levels (NOELs) taking into account all endpoints measured, but also from No Observed Adverse Effect Levels (NOAELs) that include only the higher-level endpoints like survival, weight change and feeding rate [22, 44]. Hazard quotients (HQ) were calculated applying a safety factor of 100 to NO(A)ELs to derive a reference dose (RfD). HQ, defined as the estimated daily dose divided by the

Table 6.2 Risk assessment of imidacloprid and diazinon for isopod *Porcellio scaber* [22, 23]

	Imidacloprid	Diazinon
PEC (μg/g)	0.1–0.2[a]	7–80[b]
Estimated daily dose[c]	0.008	3.2
PEL/day (μg/g wet wt./day)	10	50
LOAEC isopods[d] (μg/g food)	10	50
LOAEL isopods[d] (μg/g wet wt./day)	0.24	2.54
NOEL isopods[d] (μg/g wet wt./day)	0.24	<0.23
NOAEL isopods[d] (μg/g wet wt./day)	<0.24	0.49
RQ = PEC/LOAEC	0.02	1.6
RfD-E = NOEL/100[e] (μg/g wet wt./day)	0.0024	<0.0023
RfD-A = NOAEL/100[e] (μg/g wet wt./day)	<0.0024	0.0049
HQ: estimated daily dose/RfD-A	>3.33	653

Predicted environmental concentrations (PEC) and exposure levels (PEL) are estimated from normal recommended dosages. LO(A)EC, LO(A)EL and NO(A)EL refer to lowest or no-observable (adverse) effects concentrations or levels, with adverse being used to indicate reduction of higher-level endpoints (survival, weight gain, feeding rate); NOEL also includes lower-level biomarker, like reduction of acetyl cholinesterase (diazinon) or glutathione S-transferase (imidacloprid) activity. *RQ* risk quotient, *RfD-E* reference dose based on effects on all endpoints, *RfD* reference dose based on adverse effects on higher-level endpoints

[a] Anatra-Cordone and Durkin [45]

[b] EPA [46]

[c] Assuming consumption rate of 0.04 g/g wt. b.w./day

[d] Juveniles and adults

[e] Assessment factor of 100, including two factors of 10 for interspecies and intra-species variability, respectively

RfD, is much higher for diazinon (HQ = 653) than for imidacloprid (HQ = >3.33) (Table 6.2). This suggests that isopods are more sensitive to IMI than to diazinon. Since there are not enough data on environmentally relevant concentrations of both pesticides, both RQ and HQ values have to be considered of limited value.

The RfD is a benchmark dose operationally derived from the NOAEL by consistent application of generally order-of-magnitude uncertainty factors (UFs) that reflect various types of data sets used to estimate RfDs. For example, a valid chronic animal NOAEL is normally divided by an UF of 100. The RfD is a benchmark dose operationally derived from the NOAEL by consistent application of generally order-of-magnitude uncertainty factors (UFs) that reflect various types of data sets used to estimate RfDs. For example, a valid chronic animal NOAEL is normally divided by an UF of 100.

For our knowledge, no other ecotoxicity test exist which could provide such a complete list of toxicity data that enable comparison among all available toxicity data on a chemical of interest.

6.13 Discussion

Toxicity testing can predict toxic potential of compounds before release into the environment or asses the relative toxicities of compounds already present in contaminated areas. Accurate assessment of the toxicities of chemicals requires proper test organisms with proper toxicological endpoints and proper exposure scenario so that the bioavailable concentrations of chemicals are known or can be calculated. All these data are finally aimed to contribute in ERA.

Last 20 years terrestrial isopods are one of the most promising organisms in terrestrial ecotoxicology, but due to the fact that this test is not standardised each laboratory creates its own test design what hinders straightforward comparison among different toxicity data and subsequent use of data for ERA. Standardisation of test animals, endpoints, experimental set-up, substrate and a reference chemical is also needed.

The duration of the test procedure is related to the selected toxicological end-point. Biochemical, cytological and histological biomarkers respond within days, food consumption within weeks, whereas moult cycle and reproduction respond within months. However, in routine test the duration of the exposures should be standardised/unified and in cases of other endpoints (shorter or longer) a parallel tests of modified duration conducted.

If terrestrial isopod toxicity tests are standardised for international use, then in some cases, other species of terrestrial isopods might be used instead of *P. scaber*. For example, this substitution may be appropriate in areas where other species of terrestrial isopods are more common and ecologically significant. Future development of toxicity testing with terrestrial isopods will probably also be focused on the use of isopods that have shorter life cycles and that are parthenogenetic (e.g. *Porcellionides pruinosus* or *Trichoniscus pusillus*).

Despite their many shortcomings, laboratory-conducted single-species toxicity tests still have a central position in chemical risk assessment. After development of toxicity tests based on biochemical, physiological, behavioural, or histological end-points they could replace or complement the traditional survival, growth, and reproduction tests. Advantages of sub-organismic toxicological measurements are that they provide data on toxicity of contaminants at lower levels of biological organisation which are presumably more sensitive. Standardised toxicity tests with invertebrates, are the only one possible to generate large data bases for in silico approach. Namely in the future work with vertebrates will be limited to absolute minimum. The future lies in combination of responses of an organism to a stressor. On the horizon are new tools which will undoubtedly shape the future of single species dose response tests, these are "omic" tools. But without organism and sub-organismic responses they could not demonstrate their potential.

References

1. Escher BI, Bramaz N, Mueller JF, Quayle P, Rutishausera S, Vermeirssena ELM (2008) Toxic equivalent concentrations (TEQs) for baseline toxicity and specific modes of action as a tool to improve interpretation of ecotoxicity testing of environmental samples. J Environ Monit 10:612–621
2. Mitchell EJAK, Burgess JE, Stuetz RM (2002) Developments in ecotoxicity testing. Rev Environ Sci Biotechnol 1:169–198
3. www.nanowerk.com. Accessed 20 Sept 2014
4. Bavcon Kralj M, Černigoj U, Franko M, Trebše P (2007) Comparison of photocatalysis and photolysis of malathion, isomalathion, malaoxon, and commercial malathion-products and toxicity studies. Water Res 41:4504–4514
5. Kolpin DW, Battaglin WA, Conn KE, Furlong ET, Glassmeyer ST, Kalkhoff SJ, Meyer MT, Schnoebelen DJ (2009) Handbook of environmental chemistry. Transform Prod Synth Chem Environ 2P:83–100
6. Tennekes H (ed) (2010) The systemic insecticides: a disaster in the making. Northern Bee Books, Mytholmroyd
7. Tomizawa M, Casida JE (2003) Selective toxicity of neonicotinoids attributable to specificity of insect and mammalian nicotinic receptors. Annu Rev Entomol 48:339–364
8. Cordova D, Benner EA, Sacher MD, Rauh JJ, Sopa JS, Lahm GP, Selby TP, Stevenson TM, Flexner L, Gutteridge S, Rhoades DF, Wu L, Smith RM, Tao Y (2006) Anthranilic diamides: a new class of insecticides with a novel mode of action, ryanodine receptor activation. Pestic Biochem Physiol 84:196–214
9. www.nanovalid.eu. Accessed: 20 Sept 2014
10. Persoone G, Janssen C, De Coe W (1999) New microbiotests for routine toxicity screening and biomonitoring. Kluwer/Plenium Press, New York. ISBN 978-1-4615-4289-6
11. EC (2008) Council directive 2008/105/EC on Environmental Quality Standards (EQSs) in the field of water policy, amending and subsequently repealing council directives 82/176/EEC, 83/513/EEC, 84/156/EEC, 84/491/EEC, 86/280/EEC and amending directive 2000/60/EC of the European parliament and the council. Off J Eur Communities L348:84–97, 24 Dec 2008
12. Barcelona Convention (1976) http://www.unepmap.org/index.php?module=content2&catid=001001004
13. Rand GM, Wells PG, Mc Carty LS (1995) Introduction to aquatic toxicology. In: Rand GM (ed) Fundamentals in aquatic toxicology. Taylor and Francis, Washington, DC
14. Hopkin SP (1989) Ecophysiology of metals in terrestrial invertebrates. Elsevier Applied Science, London, p 366. ISBN 1-85166-312-6
15. Løkke H, van Gestel CAM (1998) Handbook of soil invertebrate toxicity tests. Wiley, Chichester
16. www.ecetoc.org. Accessed 20 Sept 2014
17. Drobne D (1997) Terrestrial isopods – a good choice for toxicity testing of pollutants in the terrestrial environment. Environ Toxicol Chem 16:1159–1164
18. Warburg MR (1993) Evolutionary biology of land isopods. Springer, Berlin
19. Paoletti MG, Hassall M (1999) Woodlice (Isopoda: Oniscidea): their potential for assessing sustainability and use as bioindicators. Agric Ecosyst Environ 74(1–3):157–165
20. Kammenga JE, Dallinger R, Donker MH, Kohler H-R, Simonsen V, Triebskorn R, Weeks JM (2000) Biomarkers in terrestrial invertebrates for ecotoxicological soil risk assessment. Rev Environ Toxicol Chem 164:93–147
21. Roberts AP, Oris JT (2004) Multiple biomarker response in rainbow trout during exposure to hexavalent chromium. Comp Biochem Physiol 138C:221–228
22. Drobne D, Blazic M, Van Gestel CAM, Leser V, Zidar P, Jemec A, Trebse P (2008) Toxicity of imidacloprid to the terrestrial isopod *Porcellio scaber* (Isopoda, crustacea). Chemosphere 71:1326–1334
23. Stanek K, Drobne D, Trebse P (2006) Linkage of biomarkers along levels of biological complexity in juvenile and adult diazinon fed terrestrial isopod (*Porcellio scaber*, Isopoda, Crustacea). Chemosphere 64:1745–1752

24. Stanek K, Gabrijelcic E, Drobne D, Trebse P (2003) Inhibition of acetylcholinesterase activity in the terrestrial isopod *Porcellio scaber* as a biomarker of organophosphorus compounds in food. Arh Hig Rada Toksikol 54(3):183–188
25. Drobne D, Hopkin SP (1994) Ecotoxicological laboratory test for assessing the effects of chemicals on terrestrial isopods. Bull Environ Contam Toxicol 53:390–397
26. Loureiro S, Sampaio A, Brandão A, Nogueira AJA, Soares AMVM (2006) Feeding behaviour of the terrestrial isopod *Porcellionides pruinosus* Brandt, 1833 (Crustacea, Isopoda) in response to changes in food quality and contamination. Sci Total Environ 369(1-3):119–128
27. Zidar P, Drobne D, trus J, Van Gestel CAM, Donker M (2004) Food selection as a means of Cu intake reduction in the terrestrial isopod *Porcellio scaber* (Crustacea, Isopoda). Appl Soil Ecol 25:257–265
28. Jemec A, Tišler T, Drobne D, Sepčić K, Fournier D, Trebše P (2007) Comparative toxicity of imidacloprid, of its commercial liquid formulation and of diazinon to a non-target arthropod, the microcrustacean *Daphnia magna*. Chemosphere 68(8):1408–1418
29. Drobne D, Hopkin SP (1995) The toxicity of zinc to terrestrial isopods in a standard laboratory test. Ecotoxicol Environ Saf 31:1–6
30. Drobne D, Jemec A, Pipan Tkalec Ž (2009) In vivo screening to determine hazards of nanoparticles: nanosized TiO2. Environ Pollut 157:1157–1164
31. Jemec A, Drobne D, Remskar M, Sepcic K, Tisler T (2008) Effects of ingested nanosized titanium dioxide on terrestrial isopods *Porcellio scaber*. Environ Toxicol Chem 27:1904–1914
32. Drobne D, Štrus J (1996) The effect of Zn on the digestive gland epithelium of *Porcellio scaber* (Isopoda, Crustacea). Pflugers Arch 431:247–248
33. Drobne D, Drobne S (2005) Application of computer microscopy for histopathology in isopod toxicity studies. In: Ostrander GK (ed) Techniques in aquatic toxicology, vol 2. Taylor & Francis, Boca Raton, pp 137–146 (Chapter 7)
34. Lapanje A, Rupnik M, Drobne D (2007) Gut bacterial community structure (*Porcellio scaber*, Isopoda, Crustacea) as a measure of community level response to long-term and short-term metal pollution. Environ Toxicol Chem 26(4):755–763
35. Lešer V, Drobne D, Vilhar B, Kladnik A, Žnidaršič N, Štrus J (2008) Epithelial thickness and lipid droplets in the hepatopancreas of *Porcellio scaber* (crustacea: isopoda) in different physiological conditions. Zoology 6(111):419–432
36. Žnidaršič N, Štrus J, Drobne D (2003) Ultrastructural alterations of the he-patopancreas in *Porcellio scaber* under stress. Environ Toxicol Pharmacol 13:161–174
37. Lapanje A, Drobne D, Nolde N, Valant J, Muscet B, Leser V, Rupnik M (2008) Long-term Hg pollution induced Hg tolerance in the terrestrial isopod *Porcellio scaber* (Isopoda, Crustacea). Environ Pollut 153:537–547
38. Nolde N, Drobne D, Valant J, Padovan I, Horvat M (2006) Lysosomal membrane stability in laboratory- and field-exposed terrestrial isopods *Porcellio scaber* (Isopoda, Crustacea). Environ Toxicol Chem 25:263–271
39. Fischer E, Farkas S, Hornung E, Past T (1997) Sublethal effects on an organophosphorus insecticide, dimethoate, on the isopod *Porcellio scaber* Latr. Comp Biochem Physiol 116C(2):161–166
40. Ribeiro S, Guilhermino L, Sousa JP, Soares AMVM (1999) Novel bioassay based on acetylcholinesterase and lactate dehydrogenase activities to evaluate the toxicity of chemicals to soil isopods. Ecotoxicol Environ Saf 44:287–293
41. Sousa JP, Loureiro S, Pieper S, Frost M, Kratz W, Nogueira AJA, Soares AMVM (2000) Soil and plant diet exposure routes and toxicokinetics of lindane in a terrestrial isopod. Environ Toxicol Chem 19(10):2557–2563
42. Engenheiro EL, Hankard PK, Sousa JP, Lemos MF, Weeks JM, Soares AMVM (2005) Influence of dimethoate on acetylcholinesterase activity and locomotor function in terrestrial isopods. Environ Toxicol Chem 24(3):603–609
43. Vink K, van Straalen NM (1999) Effects of benomyl and diazinon on isopod-mediated leaf soil litter decomposition in microcosms. Pedobiologia 43(4):345–359

44. Tišler T, Jemec A, Mozetič Vodopivec B, Trebše P (2009) Hazard identification of imidacloprid to aquatic environment. Chemosphere 76(7):907–914
45. Anatra-Cordone M, Durkin P (2005) Imidacloprid. Human health assessment and ecological risk assessment – final report. Syracuse Environmental Research Associates, New York, SERA TR 05-43-24-03a, 28 Dec 2005
46. EPA (2004) Interim registration eligibility decision. Prevention pesticides and toxic substances (7508C). 738-R-04-006

Chapter 7
Nanoparticles and Marine Environment: An Overview

I. Ciglenečki and V. Svetličić

Abstract In natural waters, including seawater, which can be considered as extreme environment regarding high ionic strength conditions, NPs can transform, adopting strikingly different behavior. There they might be unstable, and subject to fast aggregation and sedimentation that eliminates them from the water column. Adversely, the interaction between metal bearing NPs and organic material (OM) could change their physico-chemical properties, distribution, and persistence in the water column. In general, understanding the behavior and fate of NPs in the aquatic environment is still limited due to lack of efficient methods for their characterization.

Electroanalytical methods in combination with the state-of-the art technique (e.g. AFM) are recognized as a good choice for studying different biogeochemical processes in the marine environment, especially those related with OM, sulfur species and trace metals cycling, and the interaction and distribution between dissolved and colloidal phases. Long-term studies of surface-active particles in the northern Adriatic Sea provided evidence that biotic, as well as abiotic, transformation of OM at the micro and nanoscale are at the root of macroscopic phenomena in the sea. The advent of AFM opened a possibility to directly explore these processes at the scale that determines the fate of OM and its interaction with metal bearing NPs in the seawater.

7.1 Introduction

Typically, nanoparticles (NPs) are defined as the particles with at least one dimension within the range of 1–100 nm [1]. In natural aquatic systems, colloids are usually defined as material with one dimension between 1 nm and 1 μm, while particles are greater than 1 μm [2, 3]. The unique physicochemical properties of both colloids and NPs, particularly the high surface-to- volume ratio

I. Ciglenečki (✉) • V. Svetličić
Division for Marine and Environmental Research, Rudjer Bošković Institute, Zagreb, Croatia
e-mail: irena@irb.hr

T.A. Camesano (ed.), *Nanotechnology to Aid Chemical and Biological Defense*,
NATO Science for Peace and Security Series A: Chemistry and Biology,
DOI 10.1007/978-94-017-7218-1_7

(influencing solubility, chemical reactivity, and catalytic activity) and quantum effects (influencing colour, magnetism, hardness, and electronic properties) make them important drivers of innovation with the potential to benefit the world's entire population. These materials are now used in a wide range of manufactured industrial materials [4–6]. As examples, titania NPs are used in sunscreens, fullerene are used in tires, video screens, cosmetics, and silica NPs are used to produce solids lubricants. Proteins based NPs are used in soaps, shampoos, and detergents. More recently, metal NPs are used in environmental remediation, water and soil decontamination, [7, 8] and in medicine as a molecular tool and method for target drug delivery [9]. Inevitably, the production, use, and disposal of NPs lead to their appearance in air, waters, soils, and organisms. When released in the environment, the NPs are involved in biogeochemical cycles and little is known about their fate, interaction, and transport, or their potential risk and toxicity [10, 11]. Unfortunately, the characterization and quantification of natural and engineered NPs (ENPs) is a difficult analytical challenge since measurements are highly dependent upon particle size and nature; sample concentration and solution physiochemistry; and the fundamental principles of the analytical technique [12]. Although recently behind state-of-the art analytical techniques such as transmission electron microscopy (TEM), Atomic Force Microscopy (AFM), dynamic light scattering (DLS), fluorescence correlation spectroscopy (FCS), and flow field flow fractionation (FFFF)), several novel methods have been developed for characterization of both natural and engineered NPs. These include quartz crystal microbalance (QCM) ([13, 14] and references therein), NPs tracking analysis [12, 15], chronopotentciometry [16], voltammetry [17–23], and amperometry ([24–31] and references therein). However, no "ideal" technique was found in an environmental context.

In natural waters, including seawater, which can be considered as an extreme environment regarding high pH and ionic strength conditions, NPs could transform adopting strikingly different behavior [3, 6, 32]. There they might be unstable, or subject to fast aggregation and sedimentation that eliminates them from the water column. On the other side, the interaction between metal bearing NPs and natural organic material (NOM) could change their physico-chemical properties, distribution, and persistence in the water column. Changes of physic-chemical properties of pristine NPs will also influence their analytical detectability [32]. In general, understanding the behavior and fate of NPs in the aquatic environment is still limited due to lack of efficient methods for their characterization.

In this chapter we focus on organic colloids, metal, and sulfur bearing NPs, and their interaction and behaviour in seawater conditions. A review of recent research at the Ruđer Bošković Institute aimed at characterizing electrochemical and nano- to micro- scale behaviour of the phytoplankton produced colloids and metal sulfides NPs in the seawater conditions will be given.

7.2 Metal Sulfide and Polysaccharide Based Colloids and NPs in Aquatic Environment

Among the naturally occurring NPs, chalcogenides is one example of both natural and engineered NPs (Me_nX_m, Me = metal and X = S, Se or Te), which are a highly important class in natural waters. They are considered to be important intermediates in different mineralization and biogeochemical processes ([17–20, 33–35], and references therein). In sulfidic aquatic systems, they control the mobility, speciation, and bioavailability of many trace metals (Fe, Zn, Hg, Ag, Cu), as well as stability and distribution of sulfide species in anoxic conditions [35–37]. They may play an important role as mediators of trace metals nutrition and toxicity in aquatic ecosystems. Their production mechanisms, lifetimes, and biological effects are still unknown. It is expected that chalcogenide NPs have some toxic behaviour which is not yet fully understood [38]. These unknowns point to urgent needs for development of new analytical methods for systematic studies of chalcogenide NPs in natural waters.

In the seawater, diatoms are known as main producer of colloidal mucus and other exopolymeric material [39, 40]. This colloidal mucus through processes of aggregation, coagulation and flocculation transfer from particle size to form microaggregate which may grow futher and under particular conditions, lead to various structers of macroaggregates (macroflakes, strings, clouds, gelatins) [40]. Mucilaginous algal blooms known as mucilage events were very intensive in the North Adriatic Sea during the 1990s and early 2000s [40–46]. Chemical study of these mucilaginous macroaggregate composition showed that they are mostly composed of water-soluble heteropolysaccharides and longer polymethylene chains attached (in decreasing order) to the monosaccharide units of glucose, fructose, mannose, galactose, fucose, arabinose, ribose and xylose [44]. These results suggest the primary origin of polysaccharides in Adriatic macroaggregates from phytoplankton exudates and structural polysaccharides, which are thought to be similar in composition to phytoplankton mucus [41, 42, 45].

Some authors consider mucilage aggregate to be like an anomalous accumulation of NOM produced by several marine organisms depending on the peculiar climatic conditions that support the mucilage formation [40, 47]. Anoxic stress conditions, which could provoke massive diatom lyses in bottom or anoxic microzones followed by strong interactions between NOM and sulfide, apparent as the formation of surface-active organosulfur species (sulfopolysaccharides), were listed as one of the important factors triggering the mucilage formation in the Adriatic Sea [41, 42]. Such a view supports the prediction that mucus material in the seawater originated not only from cell secretion but also from intracellular carbon storage material from liased cells [48]. Those intracellular particulate polysaccharides, which when released from cells after interaction with sulfide, could wind up in layers and filaments to form gel-like compounds [42]. In support, Zou and collaborators [49] showed that transparent exopolymer particles (TEP), which play an important role for ocean particle aggregation and floc formation [50, 51], are found to be enriched in adhesive, surface-active sulfate polysaccharides.

7.3 Electrochemical Characterization of Chalcogenide and Sulfur Based Colloids and Nanoparticles in Seawater Conditions

7.3.1 Voltammetric Measurements

Electroanalytical methods are recognized as a good choice for studying different biogeochemical processes in the marine environment, especially those related with OM [41, 42, 52, 53], sulfur species ([17–26, 33–34, 41, 42, 54, 55] and references therein) and trace metals cycling, interaction and distribution between dissolved and colloidal phases [17–20, 33–36, 56].

Voltammetric measurement, i.e. measurements where current is measured while scanning the entire voltage range of the electrode, (*i–E* response), in our case usually from 0−1,600 mV, allows for the measurements of more than one species at a given time in the same region of space. Voltammetry at a Hg drop electrode is shown as promising toll for different sulfur species characterization and determination [26]. Figure 7.1 shows voltammetric curves recorded in the North Adriatic seawater sample containing polysacharide based mucuos aggregates [42]. This sample represent environmental system with complex interaction between different natural aquatic colloids, NPs, micro- and macro- aggregates where by voltammetric measurements presence of organosulfur species was detected.

Sulfur species, including nano- and micro- particles, readily adsorb to Hg electrode due to great affinity that this surface has for any kind of reduced S ([17–26, 34, 35, 54, 55] and references therein). The electrode reaction is a two-electron reversible process involving the oxidation of mercury to mercuric ions with immediate formation of insoluble HgS:

$$HS^- + Hg^0 \leftrightarrow HgS_{adlayer} + H^+ + 2e^- \tag{7.1}$$

The dissolution, i.e. reduction of the HgS layer (right side of reaction 7.1), is often used as a measure for “free” and labile sulfur species content ($H_2S/HS^-/S^{2-}$,S^0, S_nS^{2-}, thiols, labile metal sulfide complexes and NPs) ([17, 19, 26, 33, 34, 42, 55] and references therein) (Fig. 7.1). All these compounds deposit HgS layer on the Hg electrode surface [57].

Hg electrode is shown to effectively preconcentrate metal sulfide NPs (CuS, PbS, ZnS, FeS), enabling their detection at submicromolar concentrations in cathodic scans that reduce them to amalgams and free sulfide at potentials from −0.9 to −1.35 V (vs. Ag/AgCl) [17–22, 23, 34]. In Fig. 7.2, examples of suspensions containing CuS NPs are shown. Obtained results indicate that an Hg drop exposed to stirred, unfiltered NOM rich Adriatic seawater can be successfully used as a sulfide

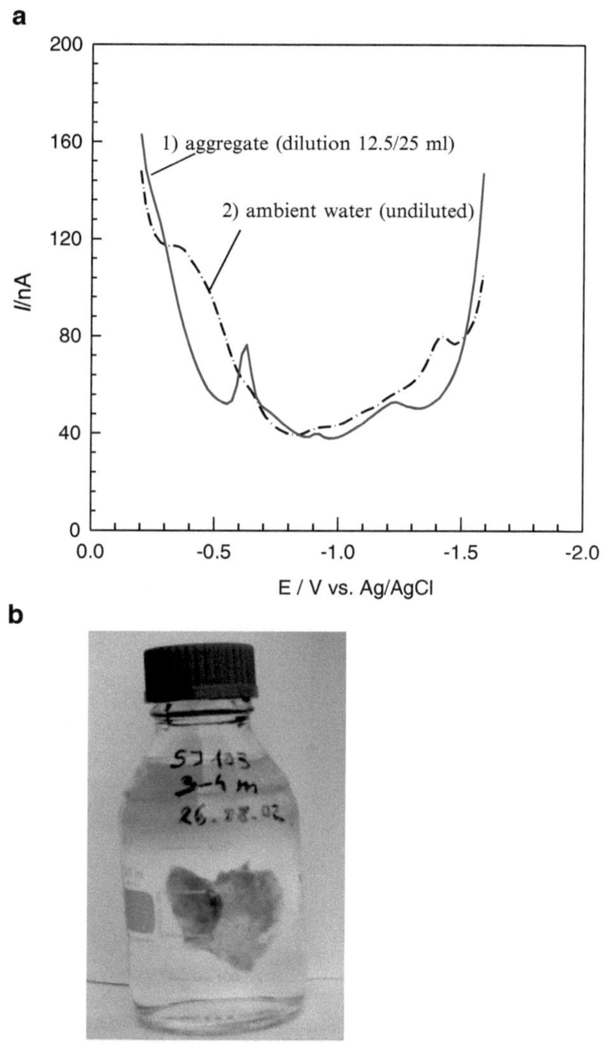

Fig. 7.1 (**a**) AC voltammograms of natural mucilage sample collected in 2002 in the North Adriatic Sea. Recorded voltammetric peaks at around −0.6 and around −1.4 V corresponds to presence of sulfur species. The first wave is ascribed to the reduction of HgS deposit formed by deposition from organosulfur species (right side of reaction 7.1) and another one is hydrogen catalytic wave produced by reduction of SH-groups from polysaccharides. These voltammetric waves were recorded only in aggregates and not in ambient waters. Their itensity were increasing during mucous ageing time. (**b**) Visulalization of polysaccharide based mucilage produced by diatoms in the North Adriatic Sea (August, 2002) (Modified from [42])

NPs collector, which can be formed in natural water environment by mixing of sulfide and trace metals rich waters [19].

Due to a relatively simple redox chemistry, not influenced by multiple relevant redox states in comparison with Cu and Fe, the PbS NPs were selected as one of the model sulfide NPs (Fig. 7.3) [26]. Likewise, in some anoxic samples, occasional voltammetric peaks at −1.2 V similar to those shown in Fig. 7.3 (peak C3) were detected. This peak was attributed to the reduction of the PbS NPs according to reaction (7.2) [26]:

$$PbS + 2e- \rightarrow Pb\,(Hg) + HS \qquad (7.2)$$

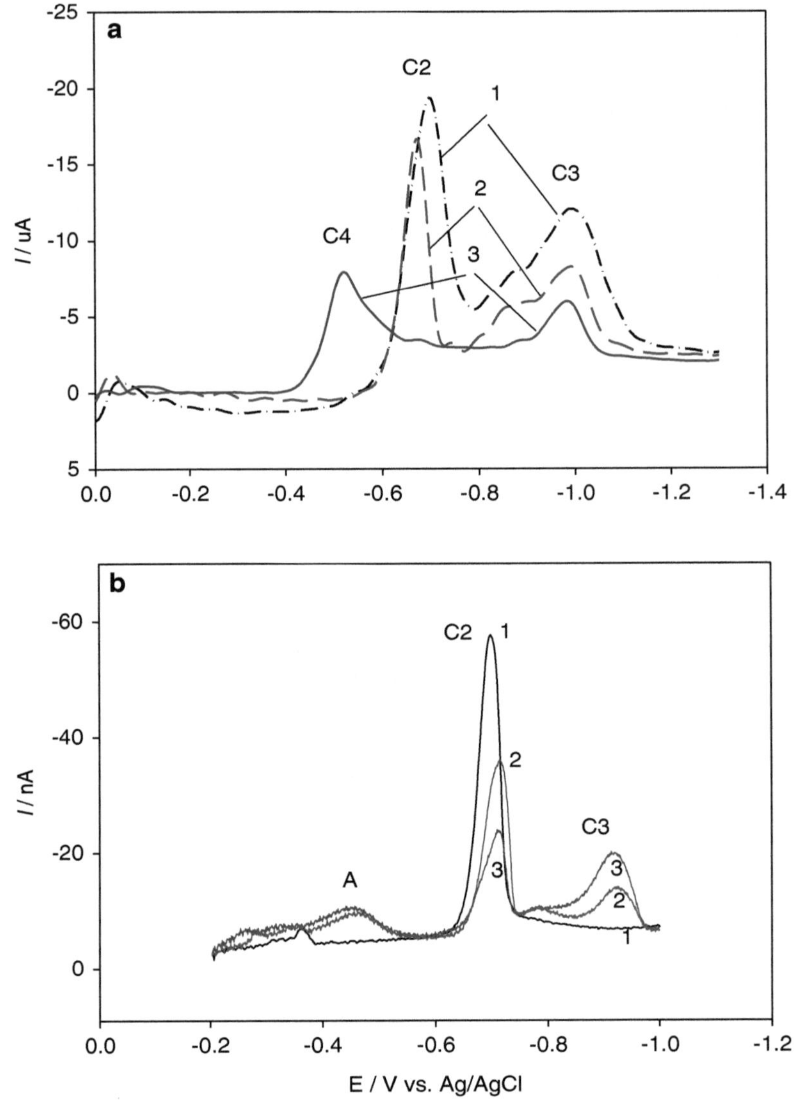

Fig. 7.2 (**a**) Linear sweep voltammograms (*LSV*) of CuS NPs grown in 1 mM $CuDCTA^{2-}$ and 1 mM HS^- in seawater conditions (0.55 M NaCl/0.03 M NaHCO3). Scan 1, initial conditions. Scans 2 and 3 after progressive purging with N2 which affected dissolution of NPs visible by diminishing of C3 and an increase of C2 and C4 peaks. Initial potential, Ei = 0 V; accumulation period $t_d = 0$ s; scan rate, v = 100 mV/s. C2 corresponds to HgS reduction (right side of reaction 7.1) deposited from free S^{2-}; C3 and C4 corresponds to reduction of CuS NPs and Cu-DCTA complex, respectively. (**b**) Voltammetric scans in mucilaginous, North Adriatic seawater that was rich in OM and contains 1.4×10^{-7} M thiols (calibrated as sulfide). Scan 1: no Cu^{2+} addition; Scan 2 and Scan 3: with 1.6×10^{-7} and 3.6×10^{-7} Cu^{2+} addition, respectively. Cu^{2+} is added in stoichiometric excess to thiol sulfur without depleting it completely. This implies that non-thiol ligands are successfully competing for some of the added Cu^{2+} on the time scale of this experiment. Initial potential, $E_i = -0.20$ V; scan rate, v = 100 mV/s; accumulation period, $t_d = 60$ s (Modified from [19])

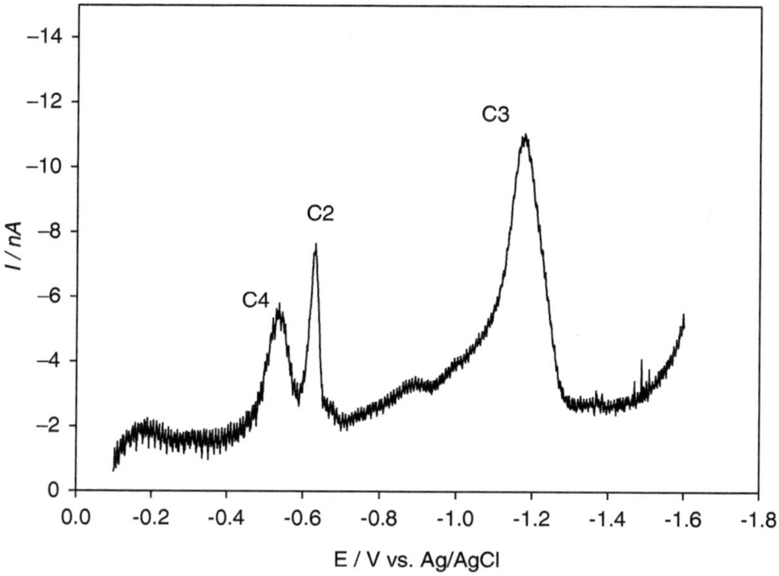

Fig. 7.3 Cyclic voltammetric scan of PbS NPs grown in seawater conditions (0.55 M NaCl/0.03M NaHCO3) containing 1×10^{-5} M Pb^{2+} and 1×10^{-5} M HS^{-}. Initial potential: $E_i = -0.2$ V; accumulation time, $t_d = 60$ s; scan rate, v = 100 mV/s. C2 corresponds to HgS reduction (right side of reaction 7.1) deposited from free S^{2-}; C3 corresponds to PbS NPs reduction; C4 corresponds to reduction of free Pb^{2+}. Initial potential, Ei = −0.1 V, scan rate, v = 100 mV/s; accumulation period, $t_d = 60$ s. All these peaks are increasing with increasing accumulation times

Direct reduction of the metal sulfide NPs [17, 24, 26], and/or reduction processes which occur on the NPs surfaces upon collision with the Hg electrode, together with the electrode potential of these processes, has been used as a background for NPs characterization in chronoamperometric measurements [24, 26].

7.3.2 *Amperometric Measurements*

Preliminary investigation of chronoamperometric signals of FeS micro- and nanoparticles at the Hg electrode [24] shows potential of the chronoamperometry for metal-bearing NPs characterization in natural waters. In comparisson with voltammetry, in chronoamperometric measurements the current (*i*) is measured as a function of time (*t*) (so called *i*-*t* curves) at a fixed potential between the working and reference electrode, and in the case of FeS, NPs revealed chronoamperometric signals at −1.9 V that appears to carry information on FeS NPs size, charge, and concentration [24]. Similar results were observed for PbS NPs at −1.5 V (Fig. 7.4) [26]. In the chloride solution containing PbS NPs, the best chronoamperometric responses were obtained in the area of the reduction peak potential marked as C3

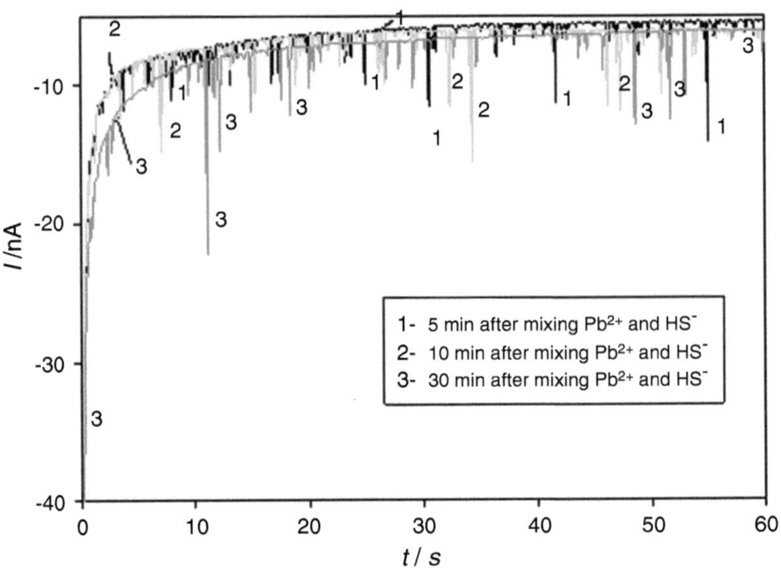

Fig. 7.4 Chronoamperograms for 5×10^{-5} M Pb^{2+} and HS^- in 0.11 M NaCl/0.03 M $NaHCO_3$ recorded at -1.5 V (vs. Ag/AgCl) 5 min (**1**), 10 min (**2**) and 30 min (**3**) after mixing of Pb^{2+} and HS^- (Modified from [26])

in Fig. 7.3. The highest frequency of spike-like signals, i.e. sharp reduction current, transients with duration lasting from 100 ms and higher and peak heights in the range of 10^{-10}–10^{-8}A were detected at a potential of -1.5 V [26]. Each peak spike signal was attributed to reduction of the single PbS NPs at the Hg surface during collision, according to reaction 7.2. Observed chronoamperometric signals appear to be sensitive on PbS NPs concentration, pH and ionic composition, and aging time. Decrease in the frequency of signals and increase of signals charge seems to be strongly influenced by aging time of the PbS suspension similarly as already obtained in the voltammetric measurements [17]. In the given experimental time shown in Fig. 7.4, size of the formed PbS NPs, monitored by DLS measurements were changed from 140 to 480 nm according to number size distribution, which influenced the feature of the recorded chronoamperometric curves [26].

7.3.3 Future Perspectives for Environmental Monitoring of NPs by Electrochemical Methods

Environmental conditions such as ionic strength, pH and water composition, content of NOM, concentration and chemical structure of NPs are expected to highly influence physico-chemical properties of the formed NPs [32], which further might effect their electrochemical responses at the Hg electrode [17, 24]. Electrochemical studies show that the frequency and size of the recorded chronoamperometric signals are related to the NPs' size, charge, and concentration, as confirmed by parallel DLS measurements. Such findings points to a great potential of the chronoamperometric measurements for the characterization, quantification and sizing of all chalcogenides and any other colloids which behave similarly at the Hg surface. In complex natural marine system it is expected that due to very peculiar environmental conditions, NPs properties and fate might be drastically changed by what for sure would influence their electrochemical behaviour. Interaction with OM might influence the surface charge of the NPs, and at the same time can cause steric shielding of NPs; both effects most probably would influence electrochemical responses. On the other hand OM might suppress electrochemical signal due to its competitive adsorption at the electrode [52, 53, 55]. Possible interferences from presence of different sulfur bearing and other NPs can be avoided by careful selection of the experimental conditions i.e. initial potential, pH and electrode materials [24]. It is already shown that FeS and PbS NPs produce spike-like signals only in the narrow potential range, allowing their distinction from each other and the colloidal S NPs [25].

7.4 AFM as a Tool in Marine Biophysics and Nanoecology

AFM connects the nanometer and micrometer length scales by utilizing a sharp probe tip that senses interatomic forces acting between the surface of a sample and the atoms at the apex of the tip. The physical basis behind AFM and its ability to "feel" the surface makes AFM a versatile tool in biophysics, allowing high resolution imaging, nanomechanical characterization, and measurements of inter- and intramolecular forces in living and non-living structures. AFM is a non-destructive method which gives real 3D images of the sample with a vertical resolution of 0.1 nm and lateral resolution of 1 nm. Measured forces range from 10^{-6} to 10^{-11} N.

Within our recent research activities, we opened a new area of research, marine biophysics, and develop tools to approach this research activity at the nanometric scale [45, 58–63]. This goes to the application of AFM platform for the marine ecology studies that will be briefly reviewed.

7.4.1 Marine Gel Phase

Long-term studies of surface-active particles in the northern Adriatic Sea provided evidence that biotic, as well as abiotic transformation of OM, at the micro-and nanoscale are at the root of macroscopic phenomena in the sea. The massive appearance of gelatinous macroaggregates (Fig. 7.5a) offered a possibility for systematic studies of the marine gel phase using AFM ([45, 62] and references therein). The samples were prepared from macroaggregates collected after different residence times in the water column, from the early stage of gel phase formation

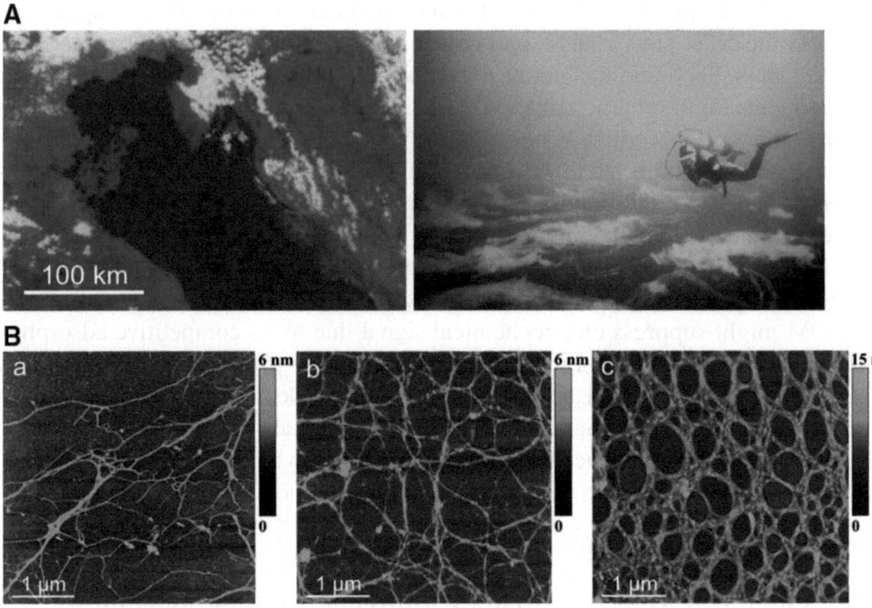

Fig. 7.5 (**A**) Northern Adriatic gel aggregates: remote sensing by satellite showing gel phase in *red* color and at 10 m depth captured by a scuba-diver. (**B**) Evolution of polymer networks in the macroscopic gel phase from early stage of gel phase formation (**a**) to condensed gel (**b**) network of older macroaggregate (**c**) (Modified from [61])

to the condensed (mature) gel network of an older macroaggregate (Fig. 7.5b). Structural details of the gel network visualized down to the molecular level revealed the monomolecular, helical and super helical associations. Our results support the hypothesis that extracellular polysaccharides produced by marine diatoms self-assemble to form a giant gel network, this process being independent on bacterial processing.

The knowledge of the mechanical strength of individual molecular assemblies within a marine gel network contributes to the understanding of the gel phase formation and its persistence in the marine environment. The physical forces in the marine gel network assemblies have been quantified using force spectroscopy together with high resolution AFM imaging [64]. Marine gels appeared highly extensible, as they could be stretched with very little force to distances of up to several micrometers. Fairly large forces are required to unzip the fibrils, suggesting that lateral stability is important in maintaining the structural integrity of the marine gel.

Such self-assembled networks with randomly distributed microscopic features and high extensibility are capable to respond to environmental conditions (such as change in salinity, temperature, pH, and shear stress), and thereby maintaining favorable physiological conditions for microbial community populating gel aggregate. From a broader perspective, our approach may be used to design experiments to study mechanical properties of other complex biopolymer networks such as the extracellular matrix in living tissues, microbial biofilms, or gel matrices for a variety of technological applications.

7.4.2 ENPs in Marine Environment: AFM Characterization of Silver Nanoparticles Interactions with Marine Diatom Cells and Extracellular Polymeric Substance (EPS)

Surprisingly, there are only a few studies on Ag NPs effect on algal cells. ENPs were found to reduce cell-specific growth rate, photosystem II quantum yield, and chlorophyll-*a* content of marine and freshwater algae [65, 66]. These toxic effects were accounted for by the release of metal ions or generation of reactive oxygen species (ROS) (e.g. [66, 67]). The toxicity of NPs themselves remains an open question, as the NP-cell wall interactions were not precisely analyzed. Lightly silicified marine diatoms *Cylindrotheca fusiformis* and congeneric *Cylindrotheca closterium* were selected for this study [58] since our previous AFM studies provided a detailed morphological and nanomechanical characterization of their cell wall and EPS release at a single cell level [46, 59, 61]. Our goal was to answer the following questions using AFM: (i) can NPs enter the diatom cell and (ii) does the exopolymeric substance produced by diatoms contribute to detoxification. We designed a set of AFM experiments in order to: (i) characterize the state of NPs in seawater in terms of size, shape and structure; (ii) visualize changes in cell morphology induced by NPs; (iii) visualize NP cell wall penetration and (iv) characterize NPs interaction with the extracellular polymeric substance.

We have found that in natural seawater used throughout this study, the single Ag NPs adopted truncated tetrahedron morphology (higest biocidal capacity, Fig. 7.6a). This size class Ag NPs penetrates the cell wall through the valve region built of silica nanoparticles embedded in organic matrix (Fig. 7.6b). The Ag NPs cause a local damage inside the cell without disintegration of the cell wall. The EPS production has been shown to increase as a feed-back response to Ag NP exposure and may contribute to detoxification mechanisms. Visualizing the extracellular polymeric substance at high resolution revealed the incorporation of Ag NPs and their aggregates into the EPS-gel matrix. The diatom EPS component responsible for the NPs interaction was identified by bringing in direct contact polysaccharide fraction isolated from *C. closterium* EPS with Ag and SiO_2 NPs (Fig. 7.7). The NPs were detected exclusively on polysaccharide fibrils as single spherical particles or their agglomerates. The NPs did not induce cross-linking of fibrils nor change the fibril heights. Rather, particles were imbedded into the preexisting polysaccharide network. Considering that EPS is a major source of dissolved and particulate organic carbon pools, possible interactions with ENPs are critically important. This nanospecific interaction could have significant environmental implications. While Ag NPs integrated in EPS gel network are beneficial to the diatom cell (detoxification), their accumulation and persistence (due to direct interaction with

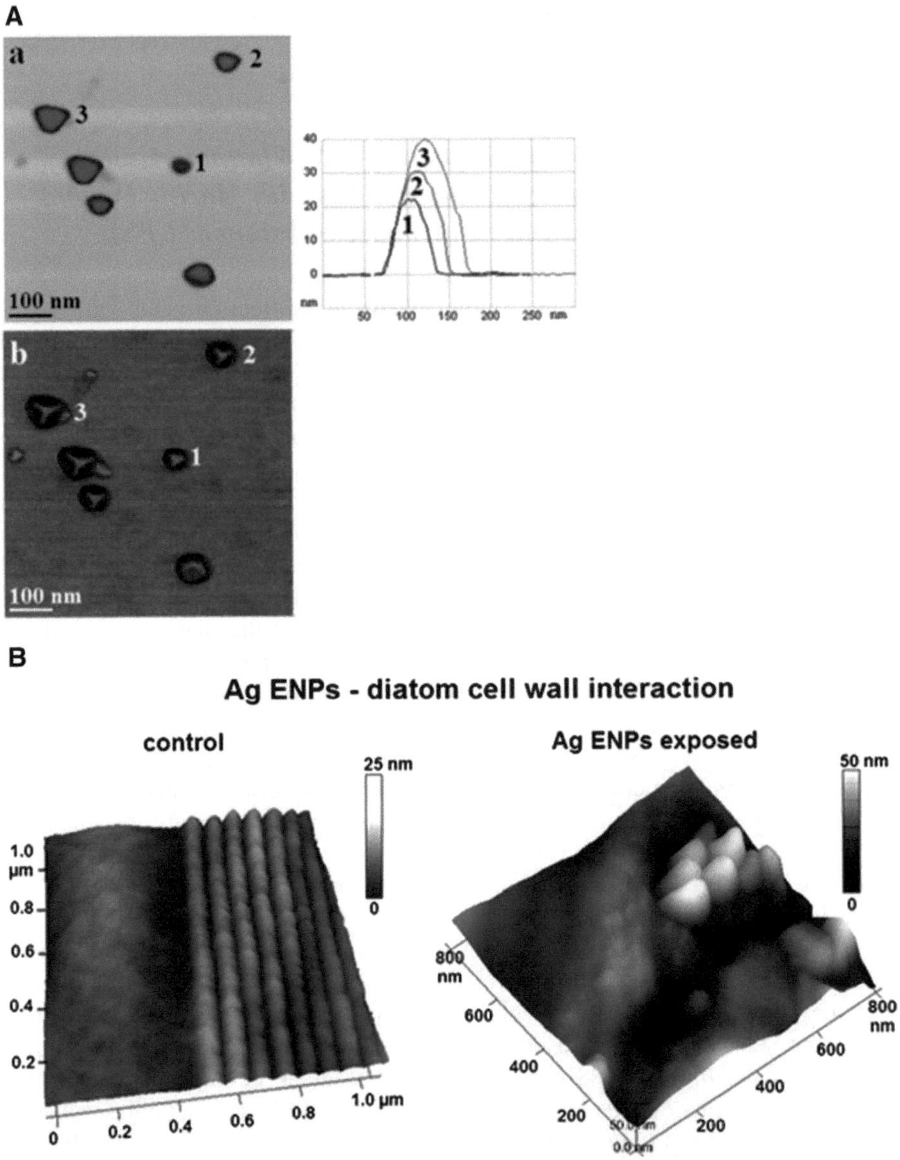

Fig. 7.6 (**A**) AFM images of individual Ag NPs in seawater: (**a**) single NPs (height image) with vertical profiles and (**b**) the corresponding phase image. (**B**) High resolution images of *C. fusiformis* cell wall prior (control) and after 24 h exposure to 10 mg/L Ag NPs. Images are presented as 3D data (Modified from [58])

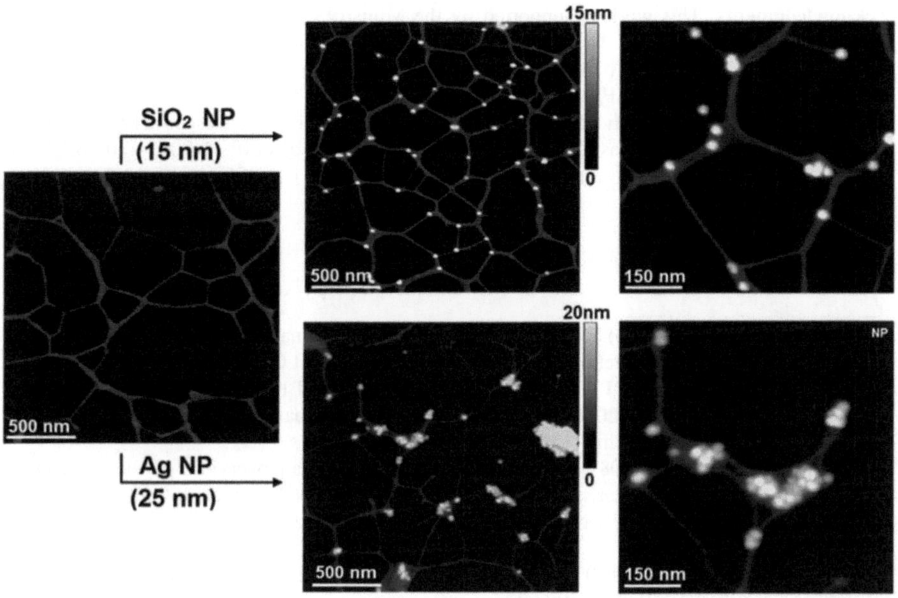

Fig. 7.7 AFM images of the self -assembled gel network sampled before and after the addition of 10 mg/L ENPs. The polysaccharide gel was prepared by dissolving the polysaccharide fraction isolated from the *C. closterium* culture medium in ultrapure water at concentration of 20 μg/mL and stirred for 45 min before adding ENPs: Ag-citrate coated, nominal size 25 nm and SiO_2 (LUDOX® HS-40, Sigma), nominal size 15 nm (Modified from [63])

the EPS polysaccharide component) in microenvironments prolong their presence in the water column and make NPs available to higher organisms. Residing in a gel environment, particles are prevented from aggregation and export from the water column. This is how Ag NPs (and most likely other engineered NPs) can persist in the water column, in spite the fact that Ag NPs quickly aggregate in seawater.

7.4.3 Final Remarks

Nawadays most environmental impact studies of nanomaterials are focused on "nanotoxicity" – the direct harmful effects on cellular level. We point out that indirect influences from EPNs can potentially pose greater environmental threats than those of direct toxicity and that AFM has potential to provide unique evidence of ENPs impact on ecosystem level.

Acknowledgments This work is supported by the Ministry of Science and Technology of the Republic of Croatia projects Nos. 098-0982934-2717 and 098-0982934-2744 and the Unity through Knowledge Fund, UKF project: "Nanoparticles in aqueous environment: electrochemical, nanogravimetric, STM and AFM studies". The COST Action TD1002: "European network on application of Atomic Force Microscopy to NanoMedicine and Life Sciences" is acknowledged for providing fruitful collaborations. Croatian Science Foundation project IP-11-2013-1205 SPHERE.

References

1. ASTM International (2006) Standard terminology relating to nanotechnology. ASTM Standard E 2456-06. American Society for Testing and Materials, West Conshohocken
2. Buffle J, Van Leeuwen H (1992) Environmental particles, vol 1. Lewis Publishers, Chelsea
3. Lead JR, Wilkinson KJ (2006) Aquatic colloids and nanoparticles: current knowledge and future trends. Environ Chem 3:159–171
4. Ju-Nam Y, Lead JR (2008) Manufactured nanoparticles: an overview of their chemistry, interactions and potential environmental implications. Sci Total Environ 400:396–414
5. Nel A, Xia T, Madler L, Li N (2006) Toxic potential of materials at the nanolevel. Science 311:622–627
6. Nowack B, Bucheli TD (2007) Occurrence, behavior and effects of nanoparticles in the environment. Environ Pollut 150:5–22
7. Kirschling TL, Gregory KB, Minkley JEG et al (2010) Impact of nanoscale zero valent iron on geochemistry and microbial populations in trichloroethylene contaminated aquifer materials. Environ Sci Technol 44(9):3474–3480
8. Ziegmann M, Frimmel FH (2010) Photocatalytic degradation of clofibric acid, carbamazepine and iomeprol using conglomerated TiO2 and activated carbon in aqueous suspension. Water Sci Technol 61(1):273–281
9. Dong Y, Feng SS (2007) In vitro and in vivo evaluation of methoxy polyethylene glycol-polylactide (MPEG-PLA) nanoparticles for small-molecule drug chemotherapy. Biomaterials 28:4154–4160
10. Linkov I, Steevens J, Adlakha-Hutcheon G et al (2009) Emerging methods and tools for environmental risk assessment, decision-making, and policy for nanomaterials: summary of NATO advanced research workshop. J Nanopart Res 11:513–527
11. Tervonen T, Linkov I, Figueira JR et al (2009) Risk-based classification system of nanomaterials. J Nanopart Res 11:757–766
12. Domingos RF, Baalousha MA, Ju-Nam Y et al (2009) Characterizing manufactured nanoparticles in the environment: multimethod determination of particle sizes. Environ Sci Technol 43:7277–7284
13. Fatisson J, Domingos RF, Wilkinson KJ et al (2009) Deposition of TiO2 nanoparticles onto silica measured using a quartz crystal microbalance with dissipation monitoring. Langmuir 25(11):6062–6069
14. Quevedo IR, Tufenkji N (2009) Influence of solution chemistry on the deposition and detachment kinetics of a CdTe quantum dot examined using a quartz crystal microbalance. Environ Sci Technol 43:3176–3182
15. Malloy A, Hole P, Carr B (2007) Nanoparticle tracking analysis; the Halo system. In: Ash B (ed) Integrated nanosensors, Mater Res Soc Symp Proc: 952E, Warrendale, 2007, 0952-F02-04
16. Pinheiro JP, Domingos R, Lopez R et al (2007) Determination of diffusion coefficients of nanoparticles and humic substances using scanning stripping chronopotentiometry. Colloids Surf A 295:200–208
17. Bura-Nakić E, Krznarić D, Jurašin D et al (2007) Voltammetric characterization of metal sulfide particles and nanoparticles in model solutions and natural waters. Anal Chim Acta 594(1):44–51

18. Bura-Nakić E, Krznarić D, Helz GR et al (2011) Characterization of iron sulfide species in model solutions by cyclic voltammetry. Revisiting an old problem. Electroanalysis 23:1376–1382
19. Ciglenečki I, Krznarić D, Helz GR (2005) Voltammetry of copper sulfide particles and nanoparticles: investigation of the cluster hypothesis. Environ Sci Technol 39(19):7492–7498
20. Helz GR, Ciglenečki I, Krznarić D et al (2011) Voltammetry of sulfide nanoparticles and the FeS(aq) problem. In: Tratnyek PG, Grundl TJ, Haderlein SB (eds) Aquatic redox chemistry. American Chemical Society, Washington, DC, pp 265–282
21. Krznarić D, Helz GR, Ciglenečki I (2006) Prospect of determining copper sulfide nanoparticles by voltammetry: a potential artifact in supersaturated solution. J Electroanal Chem 590:207–214
22. Krznarić D, Helz GR, Bura-Nakić E et al (2008) Accumulation mechanism for metal chalcogenide nanoparticles at Hg electrodes: Cu sulfide example. Anal Chem 80(3):742–749
23. Krznarić D, Ciglenečki I (2014) Voltammetric study of an FeS layer on a Hg electrode in supersaturated FeS chloride solution. Environ Chem 12(2):123–129
24. Bura-Nakić E, Marguš M, Milanović I et al (2014) The development of electrochemical methods for determining nanoparticles in the environment. Part II. Chronoamperometric study of FeS in sodium chloride solutions. Environ Chem 11(2):187–195
25. Bura-Nakić E, Marguš M, Jurašin D et al (2015) Chronoamperometric study of elemental sulfur (S) nanoparticles (NPs) in NaCl water solution: new methodology for S NPs sizing and detection. Geochem Trans 16:1. doi:10.1186/s12932-015-0016-2
26. Ciglenečki I, Marguš M, Bura-Nakić E, Milanović I (2014) Electroanalytical methods in characterization of sulfur species in aqueous environment. J Electrochem Sci Eng 4:155–163
27. Kovač S, Svetličić V, Žutić V (1999) Molecular adsorption vs. cell adhesion at an electrified aqueos interface. Colloids Surf A 149:481–489
28. Svetličić V, Ivošević N, Kovač S et al (2000) Charge displacement by adhesion and spreading of a cell: amperometric signals of living cells. Langmuir 16:8217–8220
29. Svetličić V, Ivošević N, Kovač S et al (2000) Charge displacement by adhesion and spreading of a cell. Bioelectrochemistry 53:79–86
30. Svetličić V, Hozić A (2002) Probing cell surface charge by scanning electrode potential. Electrophoresis 23:2080–2086
31. Xiao X, Fan FRF, Zhou J (2008) Current transients in single nanoparticle collision events. J Am Chem Soc 130:16669–16677
32. Delay M, Frimmel FH (2012) Nanoparticles in aquatic systems. Anal Bioanal Chem 402:583–592
33. Bura-Nakić E, Viollier E, Jezequel D et al (2009) Reduced sulfur and iron species in anoxic water column of meromictic crater Lake Pavin (Massif Central, France). Chem Geol 266:320–326
34. Bura-Nakić E, Viollier E, Ciglenečki I (2013) Electrochemical and colorimetric measurements show the dominant role of FeS in a permanently anoxic lake. Environ Sci Technol 43:741–749
35. Mullaugh KM, Luther GW III (2010) Spectroscopic determination of the size of cadmium sulfide nanoparticles formed under environmentally relevant conditions. J Environ Monit 12:890–897
36. Mullaugh KM, Luther GW III (2011) Growth kinetics and long-term stability of CdS nanoparticles in aqueous solution under ambient conditions. J Nanopart Res 13:393–404
37. Sukola K, Wang FY, Tessier A (2005) Metal-sulfide species in oxic waters. Anal Chim Acta 528:183–195
38. Ballou B, Lagerholm BC, Ernst LA et al (2004) Noninvasive imaging of quantum dots in mice. Bioconjug Chem 15:79–86
39. Kiørboe T, Hansen JLS (1993) Phytoplankton aggregate formation: observations of patterns and mechanisms of cell sticking and the significance of exopolymeric material. J Plankton Res 15:993–1018
40. Vollenweider RA, Rinaldi A (eds) (1995) Marine mucilage. Sci Total Environ 165 (Special issue), pp 1–235

41. Ciglenečki I, Ćosović B, Vojvodić V et al (2000) The role of reduced sulfur species in the coalescence of polysaccharides in the Adriatic sea. Mar Chem 71:233–249
42. Ciglenečki I, Plavšić M, Vojvodić V et al (2003) Mucopolysaccharide transformation by sulfide in diatom culture and natural mucilage. Mar Ecol Prog Ser 263:17–27
43. Degobbis D, Precali R, Ferrari CR et al (2005) Changes in nutrient concentrations and ratios during mucilage events in the period 1999–2002. Sci Total Environ 353:103–114
44. Kovac N, Faganeli J, Sket B et al (1998) Characterization of macroaggregates and photodegradation of their water soluble fraction. Org Geochem 29(5–7):1623–1634
45. Mišić Radić T, Svetličić V, Žutić V et al (2011) Seawater at the nanoscale: marine gel imaged by atomic force microscopy. J Mol Recognit 24:397–405
46. Pletikapić G, Mišić Radić T, Zimmermann H et al (2011) Extracellular polymer release AFM imaging of extracellular polymer release by marine diatom *Cylindrotheca closterium* (Ehrenberg) Reiman & J.C Lewin. J Mol Recognit 24:436–445
47. Mecozzi M, Acquistucci R, Di Nato V et al (2001) Characterization of mucilage aggregates in Adriatic and Tyrrhenian sea: structure similarities between mucilage samples and the insoluble fractions of marine humic substance. Chemosphere 44:709–720
48. Baldi F, Minacci A, Saliot A et al (1997) Cell lyses and release of particulate polysaccharides in extensive marine mucilage assessed by lipid biomarkers and molecular probes. Mar Ecol Prog Ser 153:45–58
49. Zhou J, Mopper K, Passow U (1998) The role of surface-active carbohydrates in the formation of transparent exopolymer particles by bubble adsorption of seawater. Limnol Oceanogr 43:1860–1871
50. Passow U (2002) Transparent exopolymer particles (TEP) in aquatic environments. Prog Oceanogr 55:287–333
51. Prieto L, Ruiz J, Echevarria F, Garcia CM, Bartual A, Galvez JA, Corzo A, Macias D (2002) Scales and processes in the aggregation of diatom blooms: high time resolution and wide size range records in a mesocosm study. Deep Sea Res I 49:1233–1253
52. Ćosović B, Kozarac Z, Frka S et al (2010) Electrochemical adsorption study of natural organic matter in marine and freshwater systems. A plea for use of mercury for scientific purposes. Electroanalysis 22(17–18):1994–2000
53. Ćosović B, Vojvodić V (1998) Voltammetric analysis of surface active substances in natural waters. Electroanalysis 10:429–434
54. Bura-Nakić E, Helz GR, Ćosović B et al (2009) Seasonal variations in reduced sulphur species in a stratified lake (Rogoznica Lake, Croatia); evidence for organic carriers of reactive sulfur. Geochim Cosmochim Acta 73:3738–3751
55. Ciglenečki I, Kodba Z, Ćosović B (1996) Sulfur species in Rogoznica Lake. Mar Chem 53:101–110
56. Buffle J, Tercier-Weber M-L (2005) Voltammetric environmental trace metal analysis and speciation. From laboratory to in situ measurements. Trends Anal Chem 24:172–191
57. Milanović I, Krznarić D, Bura-Nakić E et al (2014) Deposition and dissolution of metal sulfide layers at the Hg electrode surface in seawater electrolyte conditions. Environ Chem 11(2):167–172
58. Pletikapić G, Vinković Vrček I, Žutić V et al (2012) Atomic force microscopy characterization of silver nanoparticles interactions with marine diatom cells and extracellular polymeric substance. J Mol Recognit 25:309–317
59. Pletikapić G, Berquand A, Mišić Radić T et al (2012) Quantitative nanomechanical mapping of marine diatom. J Phycol 48:174–185
60. Svetličić V, Žutić V, Hozić Zimmermann A (2005) Biophysical scenario of giant gel formation in the northern Adriatic sea. Ann N Y Acad 1048:524–527
61. Svetličić V, Žutić V, Mišić Radić T et al (2011) Polymer networks produced by marine diatoms in the northern Adriatic sea. Mar Drugs 9:666–679
62. Svetličić V, Žutić V, Pletikapić G et al (2013) Marine polysaccharide networks and diatoms at the nanometric scale. Int J Mol Sci 14:20064–20078

63. Urbani R, Sist P, Pletikapić G et al (2012) Diatom polysaccharides: extracellular production, isolation and molecular characterization. In: Karunaratn DN (ed) The complex world of polysaccharide. Intech, Rijeka, pp 345–370
64. Pletikapić G, Lannon H, Murvai U et al (2014) Self-assembly of polysaccharides gives rise to distinct mechanical signatures in marine gels. Biophys J 107:355–364
65. Miao AJ, Schwehr KA, Xu C et al (2009) The algal toxicity of silver engineered nanoparticles and detoxification by exopolymeric substances. Environ Pollut 157:3034–3041
66. Navarro E, Baun A, Behra R et al (2008) Environmental behavior and ecotoxicity of engineered nanoparticles to algae, plants, and fungi. Ecotoxicology 17:372–386
67. Miller RJ, Hunters L, Muller E et al (2010) Impacts of metal oxide nanoparticles on marine phytoplankton. Environ Sci Technol 44:7329–7334

Chapter 8
Developing Sensors Based on TiO_2 Nanotubes to Detect Explosives

Mario Boehme and Wolfgang Ensinger

Abstract Within the last decade there has been a great increase in the need of trace and ultra-trace detection of explosives. Due to the very low vapor pressure of TNT and PETN, the development of efficient and sensitive detector systems seems to be complex and expensive. But the development of those systems is highly desirable, as there are considerable security needs, especially with increased use of explosives in terrorist attacks and the present surge of international terrorism.

The reported approach offers a novel procedure based on metal oxide Nanotubes that are inexpensive, lightweight, easily made, and produce cost effective devices to detect PETN.

The development will be described as a step by step procedure to produce a sensing chip device, beginning with the synthesis of the starting materials, to the supersensitive measurements of PETN explosive. As a result, the whole process is actually one of the most cost-effective methods to produce explosive sensing devices reported to date. The achieved chemical nose assembly will be able to detect PETN explosive down to $\sim$112 ppt. Besides its low detection limit, the sensing device has further advantages founded in the easy and competitive design. The explosives sensor is architecturally designed to be smaller and light weight, which will allow multifaceted application.

8.1 Introduction

Driven by the fact that explosive-based weapons have an immense damage potential and are simple and easy to deploy [1, 2], explosive-based terrorism has grown enormously during the past several years.

State of the art chemical sensors designed to detect explosive vapors are awkwardly shaped, expensive, and need improvement in regards to their limit of detection. In the fight against suicide bombers, mobile subjects, and the removal

M. Boehme (✉) • W. Ensinger
Department of Materials Science, Darmstadt University of Technology, 64287 Darmstadt, Germany
e-mail: mboehme@ca.tu-darmstadt.de

T.A. Camesano (ed.), *Nanotechnology to Aid Chemical and Biological Defense*, NATO Science for Peace and Security Series A: Chemistry and Biology,
DOI 10.1007/978-94-017-7218-1_8

of land mines [3, 4], dogs are still the most effective commercially available explosive detectors [5, 6]. Different kind of sensors used for explosive detection have been designed and engineered within the past couple of years [7–9]. Part of the major classes of energetic materials, which are used for military applications, are nitroaromatics and nitramines, e.g., 2,4,6-trinitrotoluene (TNT), hexahydro-1,3,5-triazine (RDX) and Pentaerythritol tetranitrate (PETN).

Due to the very low vapor pressure of TNT and PETN, and hence their low concentration in air, the development of efficient and sensitive detector systems seems to be complex and expensive. However, the development of those systems is highly desirable, as there are considerable security needs, especially with an increased use of explosives in terrorist attacks and the present surge of international terrorism. In between there has been miscellaneous systems and methods developed to improve explosive detection. Trisha et al. developed a fluorescence turn-on mechanism to detect high explosives RDX and PETN [10]. Anderson and colleagues demonstrated that PETN could be detected using Bead-Based Fluid Arrays [11]. Mikhaltsevitch et al. explored the detection of PETN and other explosives based on polarization enhancement of NQR signals [12].

Biomolecules have also been employed to detect PETN. For example, using an antibody-based fluorometric assay is one of many methods that is based on resonance angle changes by biomolecular interactions that occur between antigen and antibody, which are immobilized [13, 14]. Further on, several physical detection techniques to detect explosives have been developed and used. The most prominent methods are Raman spectroscopy, gas chromatography, cyclic voltammetry or high performance liquid chromatography [15]. While these systems are highly selective, some are not easily transportable or portable, and most of them are very expensive. In addition, most systems currently used require an expert operator. Regarding production costs, high sensitivity, and fast response, chemical sensing devices based on 1D nanostructures have reached strong attention in science and engineering [16–22]. Moreover, metal oxide structures seem to be promising materials for building electronic nose systems. Nanostructures offer an extremely sensitive platform to detect molecular adsorption on their surfaces [23–26]. Inorganic nanowires and nanotubes exhibit unique electrical properties that can be exploited for high sensitive detection devices based on molecular adsorption.

The vapor pressures of most common explosives at room temperature are extremely low and increase rapidly with ascending temperature [27]. Explosive vapors can easily be created by heating a sample. However, a very low vapor pressure alludes that these molecules are enormous adhesive and attempt to adsorb on surfaces very easily, especially on cooler surfaces. The adsorption will be higher on metals or metal oxides, due to their high surface energy. Therefore, metal oxide nanostructures may yield high sensitive explosive devices in the future.

In this paper, we report the preparation of high sensitive PETN sensing devices, based on TiO_2 nanotubes. The reported approach offers a novel procedure to produce small, lightweight, and cost effective detecting devices to detect PETN not reported before. The fabrication of the TiO_2 nanotube explosive sensor began with the synthesis of the TiO_2 nanotubes using a template based chemical bath deposition, according to the process which has been reported by us previously for cerium oxide nanotubes [28]. Due to its surface limitation, chemical bath deposition is a convenient and power saving manner to engender hollow nanostructures.

Semiconductive nanostructures have reached strong significance in science and engineering. For many of these materials reliable syntheses are available now and a wide range of applications in all areas of nanoscience and nanotechnology have become possible. In case of semiconductive nanomaterials, anatase titanium dioxide seems to be very interesting, as a result of its pivotal role in photochemical solar cells with high conversion efficiency caused by its large band gap from 3.2 eV corresponding to 387 nm. Due to its characteristics anatase titanium dioxide nanostructures could have novel applications in solar energy conversion, oxygen sensitivity, photo catalysis and humidity measurements [29, 30].

For deposition of anatase titanium dioxide nanostructures there have been miscellaneous methods like electro spinning, sol-gel method or chemical vapor condensation developed and used [31–33]. Fabricating metal nanostructures the template deposition method, pioneered by C.R. Martin, is one of the important processes [34, 35]. So far, various types of nanostructures obtained by electroless deposition have been successfully fabricated using chemical and physical methods [36]. However, synthesis of anatase titanium dioxide nanotubes by electroless deposition has not been reported yet.

In this announcement, we also describe an effective synthesis of anatase titanium dioxide nanotubes (TNT) with a diameter of approx. 100 nm and a wall thickness of approx. 10 nm. The anatase titanium dioxide nanotubes were grown by electroless deposition under aqueous conditions in ion track etched polycarbonate templates. Due to its surface limitation, electroless deposition is a convenient and power saving manner to engender hollow nanostructures. The structural properties of the grown anatase titanium dioxide tubes were investigated by using various analytical techniques. The growth mechanism of the anatase titanium dioxide nanotubes was proposed on the basis of experimental results.

8.2 Sample Preparation

The template consists of polycarbonate foils with a thickness of 6–30 μm. Herein we used the shelf polycarbonate membrane filters, exposed in a very controlled way to charged particles in a nuclear reactor, sold by Whatman/GE™ and comparable companies. In order to remove an undesired surface treatment of polyvinylpyrrolidone (PVP) of the commercial membrane filters, we used an additional etching process.

Therefore the off the shelf polycarbonate foils were chemically etched at 50 °C using a 6N NaOH solution containing 1 % surfactant. The resulting pore diameter is increasing linearly with etching time; the pores are of cylindrical shape. In this report we created templates with a pore diameter of approx. 50–90 nm.

8.2.1 Sensitization and Activation

Prior to the electroless deposition process, the surface of the polycarbonate template was treated with sensitization and activation solutions. In this way, the template surface becomes catalytically activated and deposition on the surface is possible. Best results are obtained using the following procedure. The sensitization was performed by an aqueous $SnCl_2$ solution containing 0.25 mol/L $SnCl_2$ and 0.3 mol/L hydrochloric acid for approximately 30 min at 45 °C. After rinsing the sensitized template with deionized water for 1 min, it was placed in the aqueous activation solution containing 0.2 mol/L $AgNO_3$ and 0.02 mol/L $Co(NO_3)_2$ a minimum of 10 min, followed by rinsing in deionized water for 1 min. To complete the activation process the template was placed for a minimum of 15 min in an aqueous solution containing 0.5 mol/L $Pd(NO_3)_2$, 0.02 mol/L $Ag_2(SO)_4$, 0.15 mol/L hydrochloric acid, and some drops of tetrafluoroboric acid at 65 °C. After rinsing with deionized water for 1 min the template was prepared for electroless deposition.

8.3 Electroless Deposition

The last step is the electroless deposition of titanium dioxide. To obtain titanium dioxide nanostructures, preferably titanium dioxide Nanotubes, the sensitized and activated template was dipped into a aqueous solution containing 0.29 mol/L $Ti_2(SO_4)_3$ and 0.8 mol/L C_2H_8BN. The used deionized water was oxygenated in advance to assure an oxygen concentration of around 13 mg/L. The temperature of the deposition bath was 2 °C to get best results. At the end of the deposition time the template with the inner grown TNTs was rinsed with deionized water for several minutes. Afterwards the TNTs were annealed at 310 °C under normal air pressure for 1 h. The time of deposition can be varied regarding the desired wall thickness. In this work the deposition time was 12 min with the result of approx. 10 nm wall thicknesses.

8.4 Resulting Nanotubes

Using polycarbonate templates results in deposition of the inner walls and the template surface. To use the template as feedstock for titanium dioxide nanotube arrays, a minimum of one side of the coated surface has to be removed. To achieve bulk titanium dioxide nanotubes, both sides of the coated template have to be removed before dissolving the template. We found the best way to remove the coated surface without damaging the fabricated tubes is to use adhesive tape. Figure 8.1a shows the surface of the adhesive tape – previously the adhesive side – with the removed template surface deposit sticking on the adhesive.

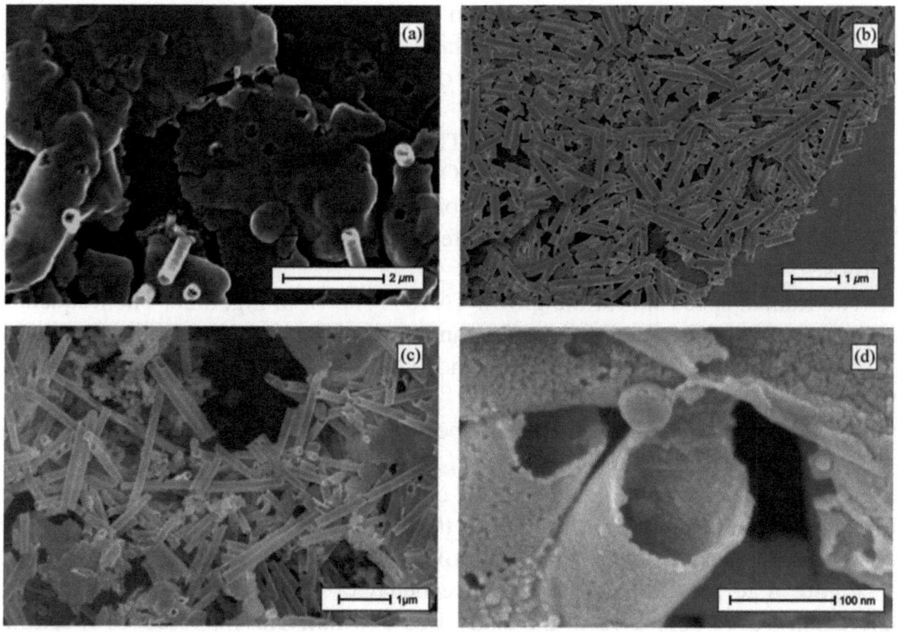

Fig. 8.1 (**a**) SEM image of the removed TiO_2 layer from the template surface. (**b–d**) SEM image of bulk TNTs unhinged from polycarbonate template

As shown in Fig. 8.1a, the removed titanium dioxide layer from the template surface mostly contains round nicks as a result of breaking up the nanotube/surface assembly, leaving the nanotubes back in the template. In this work, both sides have been removed before dissolution of the polycarbonate for further investigation.

The SEM images in Fig. 8.1b–c shows the titanium dioxide nanotubes unhinged from the polycarbonate template. The outer diameter of the tube shown in Fig. 8.1d is about 100 nm, which is in relation to the etched ion track pore size of the polycarbonate. Using high resolution SEM the surface of the titanium dioxide nanotubes is observed to be even and clogged with high magnification. Merely a small surface asperity is ascribed to the gold sputtering layer, essential to get SEM pictures due to the poor electrical conductivity of undoped TiO_2.

As expected, the nanotubes have open ends, indicating that the deposition of titanium dioxide solely occurred on the etched ion track walls. The tubes' wall thickness is linearly related to the electroless deposition conditions. As described before, the deposition time in this work was 12 min and seems to be the minimum time value to achieve stable and non fragmental TNTs. With shorter deposition times, only clusters formed on the tube wall surface without obtaining tubular structures.

By using the introduced deposition method, it is possible to assign an exact desired wall thickness simply by adjusting the deposition time. With short deposition time, tubes were achieved, while overstay deposition time increased the wall thickness. Nanowires with a small open core along their longitudinal axis were finally obtained. The reproducibility of the deposition method using polycarbonate templates was investigated by a survey over a significant number of examinable tubes and different deposition times.

The EDX spectra of the unhinged TNTs showed in Fig. 8.2a feature peaks corresponding to elements Si, Ti and O peaks, which depict the characteristic composition of the desired TNTs and confirms the presence of titanium dioxide. The Si peak arises from the silicon wafer the tubes were stored on for handling purposes.

The Raman spectrum for the grown TNTs is shown in Fig. 8.2b. The frequency and symmetry of the fundamental Raman active phonon modes for the fabricated TNTs are similar to the six allowed bands (1A1g + 2B1g + 3Eg) investigated by Ohsaka et al. [37].

As shown in Fig. 8.2b the Raman spectrum exhibits five peaks located at Eg 143 cm^{-1}, Eg 197 cm^{-1}, B1g 397 cm^{-1}, A1g + B1g 520 cm^{-1} and Eg 638 cm^{-1}, as well as one weak broad feature at 302 cm^{-1}, marked as *. The good accordance of the measured Raman Spectrum with the literature confirms that the fabricated Nanotubes consist of anatase titanium dioxide [37].

Figure 8.2c presents the fluorescence behavior of the fabricated TNTs. The excitation wavelength used in the study was 310 nm. The fluorescence spectrum shows two strong UV emission peaks at 407 and 464 nm as reported for anatase TiO2 nanostructures and confirms one more time the presence of anatase titanium dioxide [38].

The composition of the fabricated TNTs was further determined using XPS. The resulting full range survey spectrum is shown indicated in Fig. 8.3a. The two strongest peaks of the survey spectrum were investigated closely in additional spectra shown in Fig. 8.3b–c. They are located at binding energies of 530.8 and 459.5 eV respectively to the O (1 s) and Ti ($2p^{3/2}$) binding energies for TiO_2. Comparing the binding energies shown in Fig. 8.3a–c with literature values, the composition of the fabricated nanotubes was consistent with stochiometric TiO_2 [39, 40]. With further detailed examinations, these results are in good accordance to anatase TiO_2 single crystals [41].

8.5 Conclusion: Prepared Nanotubes

Bilateral open cylindrical anatase titanium dioxide nanotubes with controllable diameter and wall thickness were fabricated using electroless deposition based on aqueous solutions. This method can be extended to other materials [42–44]. Beside TNT as Bulkmaterial, a procedure to get free standing Nanotubes was developed by the authors using additional preparation steps [45]. The developed TNTs have a lot of potential applications. Using polycarbonate as template it is possible to produce TNT arrays within the size of the polycarbonate having a tube density from a single up to 10^9 tubes/cm^2. These arrays can be used directly for explosives detection applications.

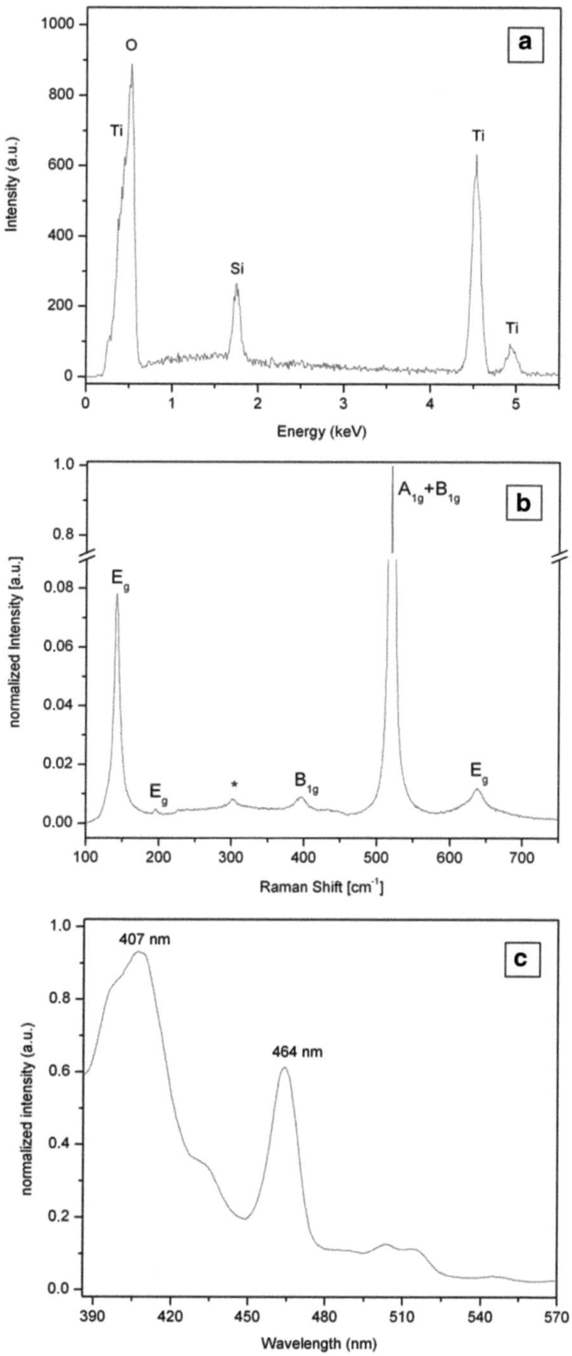

Fig. 8.2 (**a**) EDX spectrum of titanium dioxide nanotubes dissected silicon waver. (**b**) Normalized Raman Scattering spectrum of the TNTs. (**c**) Normalized PL spectra of the TNTs excited at 310 nm

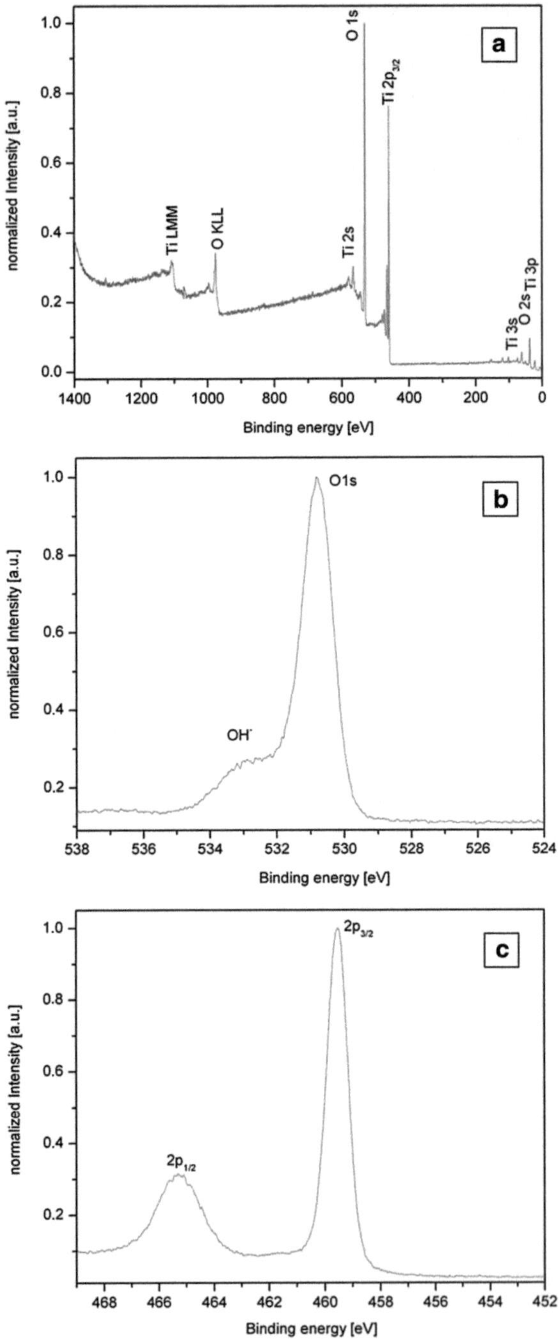

Fig. 8.3 The XPS spectra of the synthesized TNTs (**a**) full range survey spectrum; (**b**) O 1 s spectrum, (**c**) Ti 2p3/2 spectrum

8.6 Developing a PETN Sensing Device

An interdigital gold structure was used as the base of the sensing device, which was lithographically added on the surface of a Si_3N_4 coated silicon wafer. The Si_3N_4 coating had a thickness of at least 300 nm. The explosive sensing device is schematically shown in Fig. 8.4. The sensor chip consists of a meander-like structure interleaved with two terminals having a total area of 1 mm^2. Each terminal structure had two contact areas of nearly 2 mm^2. Each circuit path of the interdigital structure was 3 μm wide having a spacing of 1.5 μm to the opposite terminal meander.

To separate the nanotubes from the template, the polycarbonate foil was placed into a 2 ml vial, which was filled with dichloromethane. Further on, the vial was treated with ultrasound for 3 min. After the ultrasonic process, the vial was hydro-extracted at 13,000 rpm for 10 min. Afterwards the liquid was replaced with clean dichloromethane; taking care that the sediments, consisting of the dissolved nanotubes, retard in the vial during the removal of polycarbonate contaminated dichloromethane. The ultrasonic, hydro-extracting, and dichloromethane replacement process was repeated for ten times, to assure the TiO_2 nanotubes are free of polycarbonate. After the separation process was finished, 20 μl of the dichloromethane/nanotube suspension was taken out of the vial and carefully placed in the middle of the sensing device. Considering that the tube density of the template is well known, the concentration of nanotubes within the suspension can be varied to the sensors need.

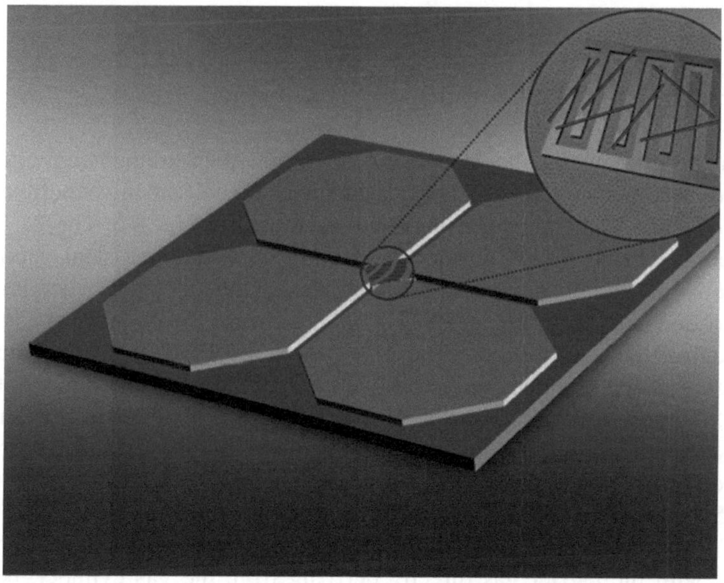

Fig. 8.4 3D-CAD drawing of the lithographic added structure onto the silicon wafer

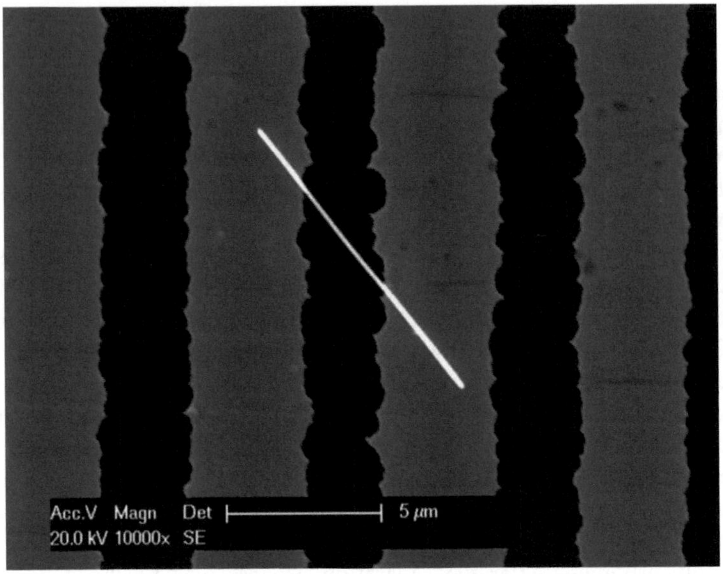

Fig. 8.5 SEM picture of sensing device showing nanotube bridging two gold terminals

After the dichloromethane was evaporated from the interdigital structure at room temperature an air ambient, the sensing chip was ready to use. A smaller area of the sensing device is shown in the SEM picture in Fig. 8.5.

8.7 Chemical Sensing Experiments

The sensing chip was connected using a 4-point measurement circuitry and a Keithly dual source meter 2412. Further on, the chemical sensing experiments were performed using a chemical sensing setup as shown in Fig. 8.6. The PETN vapor was generated from PETN powder by heating up a stainless-steel vapor source chamber having an inlet and outlet gas port. The chamber contains a ceramic sliding shuttle loaded with 100–500 mg PETN.

To control the equilibrium vapor pressure of the PETN sample material, the temperature of the chamber was controlled and set up to temperatures of 50, 60, and 70 °C during different measurements. Using argon as carrier gas the PETN vapor was delivered into the measurement chamber. The flow of the carrier gas was controlled by operation of two mass flow controllers, having an adjustable flow rate from 0.5 to 900 ml min^{-1}. To allow concentration dependence explosive sensing, the PETN vapor was easily diluted by mixing the carrier gas from the vapor source chamber with the carrier gas directly deriving from the gas cylinder, before the carrier gas reached the measurement chamber. The detailed flow rates and resulting PETN concentrations are shown in Table 8.1.

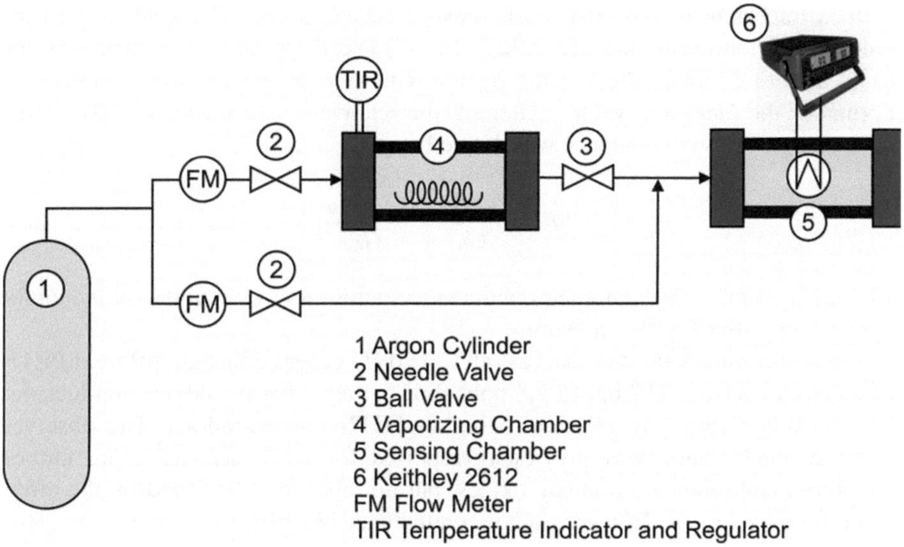

Fig. 8.6 Sensor measuring system; shown schematically

Table 8.1 Flow rate and resulting concentration of the sensing measurement

Temp. [°C]	Approx. conc. in vaporizer chamber based on [27] [ppt]	PETN/argon flow rate [ml/min]	Approx. PETN conc. at sensor-chip [ppt]
50	1400	40/500	112
60	6592	30/750	263
		30/500	393
		30/300	660
70	28,353	30/900	945
		30/450	1890
		30/300	2835

The flow rates were selected in accordance to the volume of both the vaporizer chamber and the sensing chamber to assure a constant explosive vapor concentration during each measuring cycle. Due to the behavior of explosive vapors, tending to adsorb on cooler surfaces, the piping leading to the measurement chamber was additionally heated up to the temperature of the source chamber. Prior to sensing each measurement, the adsorbed substances were removed from the sensor using ultra violet irradiation, having a wavelength of 290 nm. The device was irradiated for 3–5 min at air ambient as previously successfully demonstrated [46]. In addition the presence of titanium dioxide, which is well known for its self cleaning behavior under ultraviolet irradiation, in combination that PETN molecules are easy to decompose into small molecular fragments under light illumination [47, 48], will support the reset of the sensing device.

It is important to note that each sensing device is one of a kind, due to the random distribution of the nanotubes. To evaluate the sensing measurement and to be able to compare the sensing device with similar ones, it was necessary to normalize the measured values. Thereof, the relative sensor response (SRr) based on the assembly's conductance was defined as

$$SR_r = \frac{\Delta G}{G_0} = \frac{G - G_0}{G_0} \tag{8.1}$$

where G_0 denotes the nanotube conductance before and G denotes the nanotube conductance after PETN exposure.

Figure 8.7 shows the I–V curves (at $V_g = 0$ V) generated using different PETN concentrations from 112 ppt to 2,8 ppb, showing also that the device conductance was steadily reduced in relation to higher PETN concentrations. The observed reduced conductance was previously reported for DNT sensing using carbon nanotubes and reduced graphene oxides, but no report can be found on the effect of PETN exposure on TiO_2 nanotubes conductance [49, 50].

Figure 8.8 shows the diversification of the TiO_2 nanotubes conductance normalized to the initial conductance. Seven cycles have been recorded successfully, corresponding to seven different PETN concentrations at argon ambient, ranging

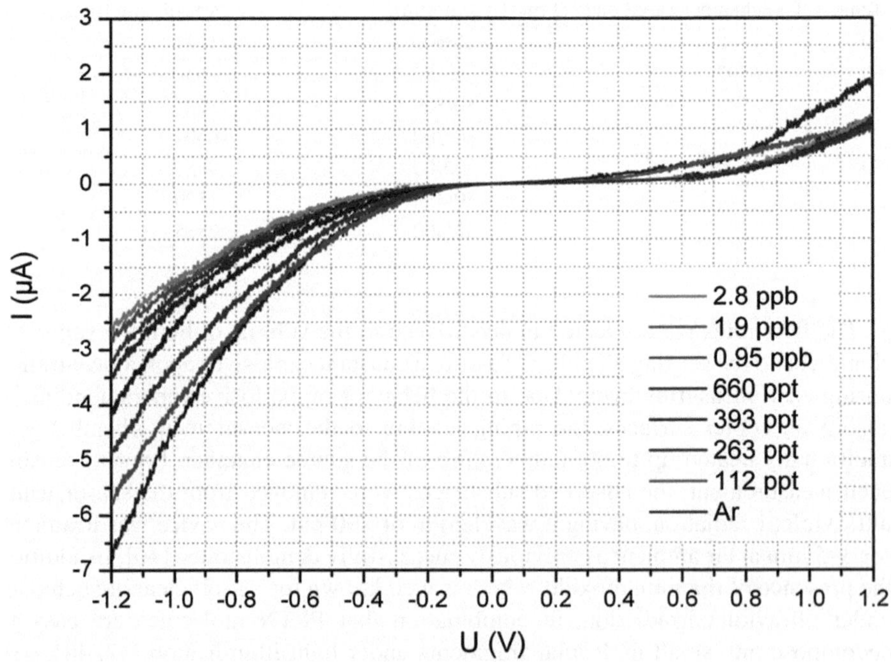

Fig. 8.7 I–V curves taken at different PETN concentrations at Argon ambient

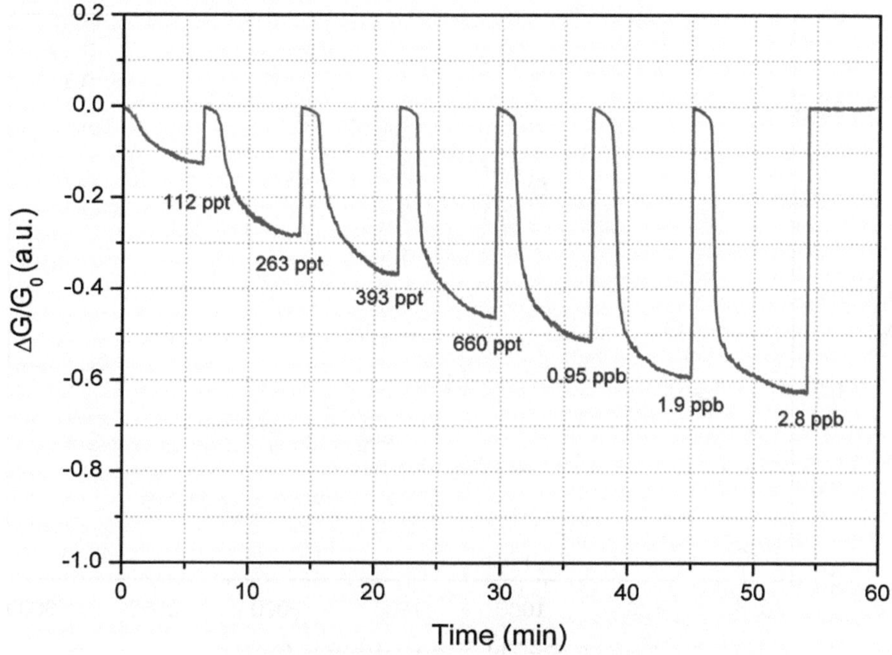

Fig. 8.8 Sensing response of the TiO_2-nanowire explosive sensing device to PETN at −1.2 V

from 112 ppt to 2,8 ppb. With increasing of the PETN concentration, more distinctive conductance suppression was observed. The lowest detectable PETN concentration was nearly 112 ppt for the described TiO_2 nanotube sensor. To our best of knowledge, this is the lowest detection limit ever reported using metal oxide nanotube based chemical sensors to detect PETN.

Figure 8.9 presents the normalized sensor response, shown in Fig. 8.8, as a function of PETN concentration. The plot has a large slope at lower concentrations descending into a plot with a smaller slope at higher concentration. It seems, that the surface coverage tends to saturate at a higher concentration, and leads to the saturation response accordingly, observed in Fig. 8.9.

Thereof, the plot can be well fitted with a Langmuir-Isotherm. Because of the very good correlation of the Langmuir-Isotherm fitting, the results of sensor measurements can be comprehended as a surface coverage of adsorbed molecules, analog to the mechanism described by Langmuir [51].

Further on, Fig. 8.9 can be understood as the calibration curve of the sensing device because of the functional relationship for the PETN which relates the expected value of the observed signal to the explosive vapor amount. The calibration curve includes important information about the sensitivity of the sensing device, which is known as the slope of the calibration curve. The higher the slope of the calibration curve, the higher is the sensitivity of the sensing device for PETN. Regarding Fig. 8.9, the developed sensing device has the highest slope at lower

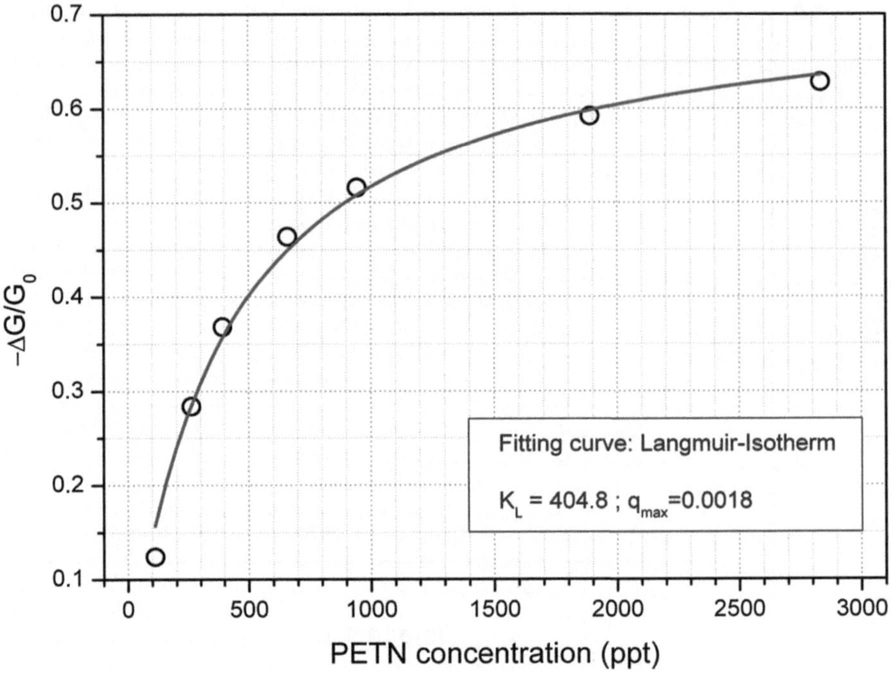

Fig. 8.9 Normalized conductance modification – $\Delta G/G_0$ vs. PETN concentration

PETN concentrations, which shows that the device is able to discriminate between small differences of the PETN amounts. In association of the calibration curve with the I–V curve, shown in Fig. 8.8, it appears that there is much space between the 112 ppt-current and the argon-current. This gap enables the sensing device to provide additional discriminated PETN sensing potential for concentrations far below the reported 112 ppt. The limitations for the sensing device in this report are founded in the facility to generate proper PETN vapors below the reported limit of detection.

8.8 Conclusion

Developing sensing devices to detect trace or ultra-trace explosives is a challenging undertaking, originated by operational factors like the marginal abundance of explosive vapors or molecules. Pursuant to our work, nanotubes are the structures of choice to build sensing devices that will satisfy the sensitivity requirements for trace explosive detection. We successfully demonstrated an easy and inexpensive to build TiO_2 nanotube-based sensing device. Our device was showing good sensitivity to PETN by having a detection limit that is far below other systems, like Airport sniffers with a detection limit of 2000 ppt, or luminescence methods which have a detection limit of 19.8 ppb at gaseous ambient [52].

Besides its low detection limit, the sensing device described in this work has further advantages founded in the easy and competitive procedure in which the device is built; the sensor is architecturally small designed and light weight, which will allow multifaceted application fields.

References

1. Hallowell SF (2001) Screening people for illicit substances: a survey of current portal technology. Talanta 54(3):447–458
2. Fainberg A (1992) Explosives detection for aviation security. Science 255:1531–1537
3. Colton RJ, Russel JN Jr (2003) COUNTERTERRORISM: making the world a safer place. Science 299(5611):2
4. Czarnik AW (1998) A sense for landmines. Nature 394(6692):417–418
5. Furton KG, Myers LJ (2001) The scientific foundation and efficacy of the use of canines as chemical detectors for explosives. Talanta 54(3):487–500
6. Hiltmar Schubert AR-K (ed) (2006) Stand-off detection of suicide bombers and mobile subjects, NATO security through science series. Springer, Dordrecht
7. Meaney M, McGuffin V (2008) Luminescence-based methods for sensing and detection of explosives. Anal Bioanal Chem 391(7):2557–2576
8. Singh S (2007) Sensors – an effective approach for the detection of explosives. J Hazard Mater 144(1–2):15–28
9. Smith RG, D'Souza N, Nicklin S (2008) A review of biosensors and biologically-inspired systems for explosives detection. Analyst 133(5):571–584
10. Andrew TL, Swager TM (2007) A fluorescence turn-on mechanism to detect high explosives RDX and PETN. J Am Chem Soc 129(23):7254–7255
11. Anderson GP et al (2010) Bead-based fluid array detection of Pentaerythritol Tetranitrate: comparison of Monoclonal vs. Llama Polyclonal antibodies. Anal Lett 43(18):2913–2922
12. Mikhaltsevitch VT, Beliakov AV (2006) Polarization enhancement of NQR signals for explosive detection. Solid State Commun 138(8):409–411
13. Judd LL et al (1995) Antibody-based fluorometric assay for detection of the explosives TNT and PETN. SPIE, San Jose
14. Hilmi A, Luong JHT, Nguyen A-L (1999) Development of electrokinetic capillary electrophoresis equipped with amperometric detection for analysis of explosive compounds. Anal Chem 71(4):873–878
15. Yinon J (2003) Peer reviewed: detection of explosives by electronic noses. Anal Chem 75(5):98 A–105 A
16. Wang ZL (2004) FUNCTIONAL OXIDE NANOBELTS: materials, properties and potential applications in nanosystems and biotechnology. Annu Rev Phys Chem 55(1):159–196
17. Law M, Goldberger J, Yang P (2004) Semiconductor nanowires and nanotubes. Annu Rev Mater Res 34(1):83–122
18. Lu JG, Chang P, Fan Z (2006) Quasi-one-dimensional metal oxide materials–synthesis, properties and applications. Mater Sci Eng: R: Rep 52(1–3):49–91
19. Snow ES, Perkins FK, Robinson JA (2006) Chemical vapor detection using single-walled carbon nanotubes. Chem Soc Rev 35(9):790–798
20. Patolsky F, Lieber CM (2005) Nanowire nanosensors. Mater Today 8(4):20–28
21. Grobert N (n.d.) Carbon nanotubes – becoming clean. Mater Today 10(1–2):28–35
22. Allen BL, Kichambare PD, Star A (2007) Carbon nanotube field-effect-transistor-based biosensors. Adv Mater 19(11):1439–1451
23. Sysoev VV et al (2007) A gradient microarray electronic nose based on percolating SnO2 nanowire sensing elements. Nano Lett 7(10):3182–3188
24. Baik JM et al (2010) Tin-oxide-nanowire-based electronic nose using heterogeneous catalysis as a functionalization strategy. ACS Nano 4(6):3117–3122

25. Cerrato Oliveros MC et al (2002) Electronic nose based on metal oxide semiconductor sensors as a fast alternative for the detection of adulteration of virgin olive oils. Anal Chim Acta 459(2):219–228
26. Po-Chiang C, Guozhen S, Chongwu Z (2008) Chemical sensors and electronic noses based on 1-D metal oxide nanostructures. Nanotechnol IEEE Trans Nanotechnol 7(6):668–682
27. Dionne BC et al (1986) Vapor pressure of explosives. J Energetic Mater 4(1):447–472
28. Boehme M et al (2011) Cerium (IV) oxide nanotubes prepared by low temperature deposition at normal pressure. Nanotechnology 22(6):065602
29. Cox A (1992) Photochemistry 22:505
30. Gonzalez RZRJ (1997) In: Thorpe MF (ed) NATO ASI proceedings
31. Xia DLY (2004) Nano Lett 4(5):933–938
32. Sadeghzadeh Attar A et al (2008) Synthesis and characterization of anatase and rutile TiO2 nanorods by template-assisted method. J Mater Sci 43(17):5924–5929
33. Maiyalagan T, Viswanathan B, Varadaraju UV (2006) Fabrication and characterization of uniform TiO2 nanotube arrays by sol–gel template method. Bull Mater Sci 29(7):705–708
34. Martin CR (1994) Nanomaterials: a membrane-based synthetic approach. Science 266(5193):1961–1966
35. Klein JD et al (1993) Electrochemical fabrication of cadmium chalcogenide microdiode arrays. Chem Mater 5(7):902–904
36. Shi Z (2006) Nanotechnology 17:2161–2166
37. Ohsaka T, Izumi F, Fujiki Y (1978) Raman spectrum of anatase, TiO2. J Raman Spectrosc 7(6):321–324
38. Fang D et al (2008) Fabrication and photoluminiscent properties of titanium oxide nanotube arrays. J Braz Chem Soc 19(6):1059–1064
39. Murata M, Wakino K, Ikeda S (1975) X-ray photoelectron spectroscopic study of perovskite titanates and related compounds: an example of the effect of polarization on chemical shifts. J Electron Spectrosc Relat Phenom 6(5):459–464
40. Gonbeau D et al (1991) XPS study of thin films of titanium oxysulfides. Surf Sci 254 (1–3):81–89
41. Silversmit G, De Doncker G, De Gryse R (2002) A mineral TiO2 (001) anatase crystal examined by XPS. Surf Sci Spectra 9(1):21–29
42. Boehme M, Ensinger W (2011) Fabrication of zinc oxide nanotubes by chemical bath deposition using ion track-etched templates. IEEE Trans Nanotechnol 10(1):63–69
43. Boehme M et al (2011) Room temperature synthesis of samarium oxide nanotubes using cost-effective electroless deposition method. J Exp Nanosci 158:286–291
44. Boehme M et al (2011) Room temperature synthesis of indium tin oxide nanotubes with high precision wall thickness by electroless deposition. Beilstein J Nanotechnol 2:119–126
45. Boehme M, Ensinger W (2011) From nanowheat to nanograss: a preparation method to achieve free standing nanostructures having a high length/diameter aspect ratio. Adv Eng Mater. In Print Accepted 20.12.10
46. Zhang D et al (2003) Ultraviolet photodetection properties of indium oxide nanowires. Appl Phys A: Mater Sci Process 77(1):163–166
47. Giefers H, Pravica M (2008) Radiation-induced decomposition of PETN and TATB under extreme conditions. J Phys Chem A 112(15):3352–3359
48. Volltrauer HN (1982) Real time low temperature decomposition of explosives – PETN. J Hazard Mater 5(4):353–357
49. Snow ES, Perkins FK (2005) Capacitance and conductance of single-walled carbon nanotubes in the presence of chemical vapors. Nano Lett 5(12):2414–2417
50. Robinson JT et al (2008) Reduced graphene oxide molecular sensors. Nano Lett 8(10):3137–3140
51. Atkins PW (2006) Physikalische Chemie. Wiley-VCH, Weinheim
52. Wilson R, Clavering C, Hutchinson A (2003) Electrochemiluminescence enzyme immunoassays for TNT and Pentaerythritol Tetranitrate. Anal Chem 75(16):4244–4249

Chapter 9
O-Doped Carbon Nitride (O-g-C_3N) with High Oxygen Content (11.1 mass%) Synthesized by Pyrolysis of Pyridine

Ganna Kharlamova, Oleksii Kharlamov, and Marina Bondarenko

Abstract Graphite-like carbon nitride (g-C_3N_4) is the most perspective material for nanosensors creation used in nanodimensional systems that aid in the chemical and biological protection of human health and environment. Carbon nitride has an extremely wide interval of region of homogeneity ($0.55 < C/N < 4.0$), which makes it an unique object because of the doping of its structure by non-metal atoms with the purpose of essential expansion of its chemical and electronic characteristics. We developed a new method of pyridine pyrolysis with the purpose of obtaining O-doped graphite-like carbon nitride (O-g-C_3N) with the high contents of oxygen (11.1 mass%). In the high-temperature (1050 °C) zone of reactor is realized mainly destruction of molecules C_5H_5N to formation of small carbon clusters (C_2—C_4) and HCN. Nanodispersed high-defected carbon nitride at the expense of polycondensation of molecules, HCN is formed in dynamic reactionary vapour-gaseous flow of all reactor, but is located only in the low-temperature (<70 °C) zone. From air valence-unsaturated atoms of carbon in resulting carbon nitride chemisorbs oxygen with formation of O-g-C_3N. In mass spectra the products of thermolysis of synthesized O-g-C_3N are detected cyanogen (C_2N_2), fragments C_4N_2 and C_2N, as well as atoms of carbon and nitrogen.

9.1 Introduction

Graphite-like carbon nitride (g-C_3N_4) is one of the most perspective materials for multifunctional non-metallic catalysts, photovoltaic solar cells and nanosensors [1, 2], and in particular for chemical and biological protection of environment and

G. Kharlamova (✉)
Taras Shevchenko, National University of Kyiv, 64 Volodymyrska str., 03001 Kyiv, Ukraine
e-mail: akharlamova@ukr.net

O. Kharlamov • M. Bondarenko
Frantsevich Institute for Problems of Materials Science of NASU, Krzhyzhanovsky St. 3, 03142 Kiev, Ukraine

T.A. Camesano (ed.), *Nanotechnology to Aid Chemical and Biological Defense*, NATO Science for Peace and Security Series A: Chemistry and Biology, DOI 10.1007/978-94-017-7218-1_9

people health. So, nanoporous carbon nitride can be used as the photocatalyst of decomposition of water and polluting substances, photofixings CO_2, as well as the catalyst for reactions of oxidation, hydrogenation [3] and photofixation of oxygen in fuel elements [4]. The use of g-C_3N_4 as photosensitizer and photocatalyst for generation of hydrogen is marked [5]. Carbon nitride received from a mix of urea and melamine (Fig. 9.1) is capable of generating hydrogen with high speed (135 mkmol/h) [6]. Due to the unique optical and electronic properties, g-C_3N_4 is applied as photo-electric nanosystems, chemical sensors, and photo-electric solar elements [7], as well as a high-sensitivity fluorescent probe for detection of ions Fe_3^+ and Cu_2^+ [2]. Porous carbon nitride received by pyrolysis of hollow spheres of polyaniline can be applied as a supercondensator [8]; the ability to preserve after 5000 cycles of charge/discharge (at density of a current 1.0 A/g) makes more than 91 % [8]. Films containing amorphous g-C_3N_4 have wear resistance, sometimes exceeding the similar characteristic of existing coverings, and therefore are used as a material for protective coverings of hard disks of computers [9]. Carbon nitride is used as non-metallic ferromagnet/paramagnet in spintronic devices [10, 11]. Depending on its composition (ratio C/N), it can be a room temperature ferromagnet [7, 10] or paramagnet [12, 13]. Its magnetism is directly connected to concentration of defects on atoms of carbon [10, 11].

The so various and unique properties of carbon nitride are caused by its main feature: g-C_3N_4 as, for example, and cubic phases of inculcation (carbides and nitrides of d-metals) has a wide interval of area of homogeneity with ratio C/N from 0.55 to 4.28 (Fig. 9.1).

As the degree of heptazine (tri-*s*-triazine) C_6N_7 fragments was increased, there was a decrease in graphite-like structure. The deviation of composition of g-C_3N_4 from its formula (C/N = 0.75) promotes to increase the defects of structure, growth of number of the active centers, and, as a consequence, improvement, such as its catalytic properties. The physic-chemical properties of samples of carbon nitride essentially depend also on morphology of nanostructures (nanospheres, nanoribbons, nanorods, nanosheets) and nanoparticles of g-C_3N_4 [1, 14, 15].

For the first time, graphite-like carbon nitride on composition close to formula one (C/N = 0.75) was obtained in 1922 by means of decomposing mercury thiocyanate $Hg(NCS)_2$ [16], and contained from 0.6 % up to 2.4 % of hydrogen. Later (after 80 years) carbon nitride of formula composition was obtained only at solid phase, with the interaction of cyanuric chloride ($C_3Cl_3N_3$) with calcium cyanamide ($CaCN_2$) in autoclave at temperature 500–550 °C [17]. Cyanamide, dicyandiamide and melamine in absence of cyanuric chloride are transformed at 520 °C to samples g-C_3N_4 with the large contents of nitrogen into the ratio C/N in which makes 0.671, 0.685, and 0.676 accordingly [18]. At interaction cyanuric chloride with lithium azide (LiN_3) [19] or sodium azide and zinc (NaN_3+Zn) [20] the samples of carbon nitride with the smaller contents of nitrogen (C/N 0.80 and 1.01 accordingly) are formed. Using such precursors as, for example, 3,3′-diaminobenzidine, triazine derivatives, the polymer HCN, diaminomaleonitrile, 1,2,4-triazole, tetracyanoethylene results in obtaining an amorphous product containing 20–45 % nitrogen [21]. It is even less nitrogen (C/N = 4) than is contained

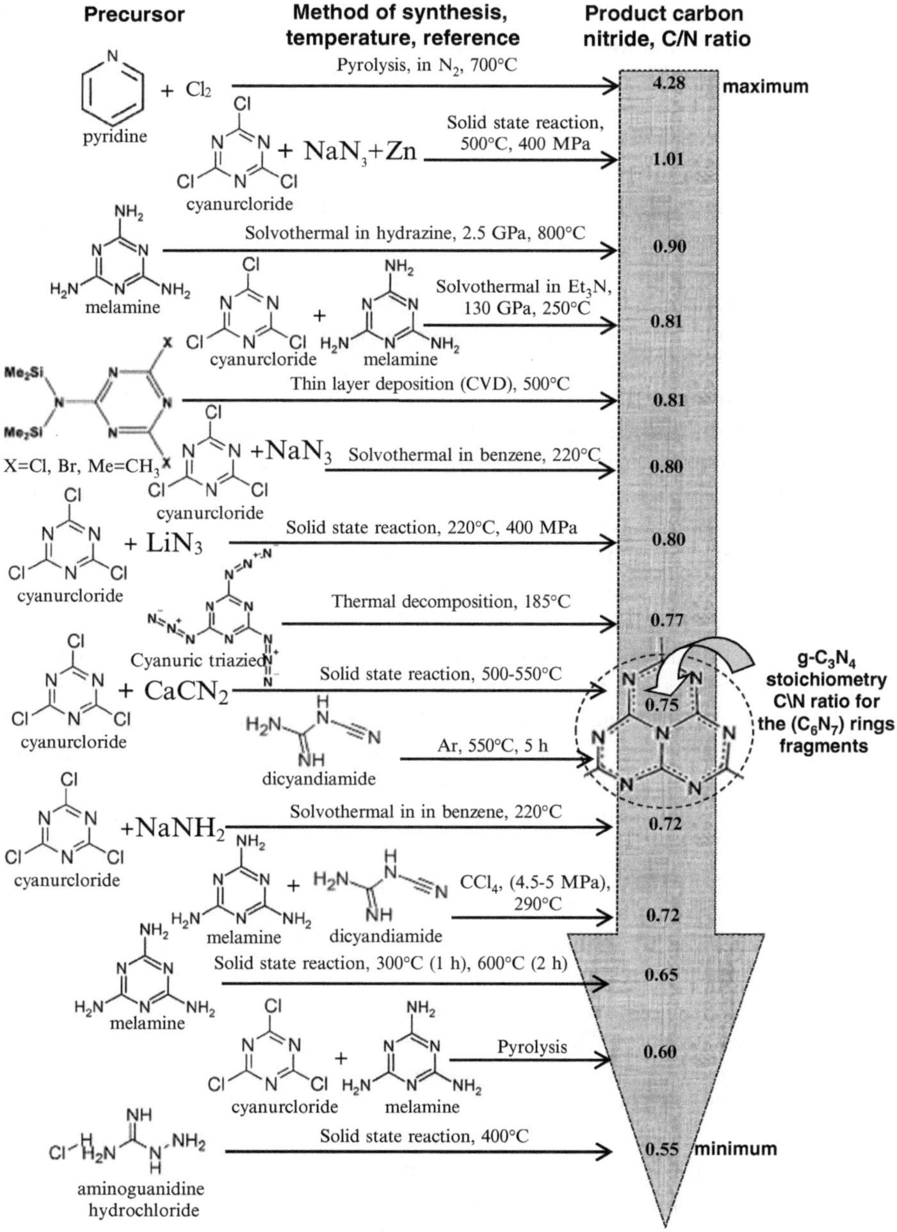

Fig. 9.1 Main methods for the carbon nitride synthesis [1, 14, 22, 23] with maximum, minimum and closest to the stoichiometric (0.75) C/N ratio

in graphite-like carbon nitride received at interaction of molecules of pyridine and chlorine ($C_5H_5N+5/2\ Cl_2 \rightarrow 7\ C_5N+5\ HCl$) at temperatures >680 °C and atmospheric pressure in quartz flowing CVD reactor [22]. Graphite-like carbon nitride with the minimal contents of nitrogen adequate to composition C_5N is formed from pyridine at use of surplus of chlorine. At a stoichiometric ratio of reagents optimum temperature of chlorination of pyridine in a flow of nitrogen makes 800 °C [24, 25], whereas in a helium flow is 700 °C [26]. In Bailey [22] is noted, that in absence of chlorine pyrolytic condensation of pyridine was not observed in general. At use as precursors such nitrogen containing compounds as piperidine, 2,6-dichloropyridine, 2,6-dicyanopyridine was obtained graphite-like nitride carbon with the very small contents of nitrogen [22]. In Talapaneni et al. [23] inform the preparation at 400 °C from aminoguanidine hydrochloride ((CH6N4)•HCl) of the product with high contents of nitrogen ($C/N = 0.55$), which, as believe the authors [23], corresponds to highly stable polymer consisting not from heptazine, but from diamino-*s*-triazine fragments. At use of such containing halogen and nitrogen compounds, as $C_3N_3F_2N(SiMe_3)_2$ and $C_3N_3Cl_2N(SiMe_3)_2$ deposited from a vapour phase of graphite-like carbon nitride contains considerably less nitrogen (C/N 0.81) [24, 25].

The quantity of nitrogen in received carbon nitride essentially depends not only on composition of precursors and their concentration, but also from a method of its synthesis. So, from a mix of melamine ($C_3H_6N_6$) and cyanuric chloride ($C_3Cl_3N_3$) receive at a pyrolytic method a product to high ($C/N = 0.60$) contents of nitrogen $C_3N_{4.91}H_{1.00}O_{0.22}$ [27], whereas at solvothermal synthesis the product with essential by the smaller (C/N 0.81) contents of nitrogen [28] is formed. Is remarkable, that the product of solvothermal synthesis, which is carried out in a solution of triethylamine (Et_3N) (or hydrazine (N_2H_4)), contains is abnormal high (~37 or ~19 at.% accordingly) quantity of hydrogen [29]. Solvothermal interaction at 220 °C cyanuric chloride in a benzene solution with sodium amide ($NaNH_2$) [30] or sodium azide (NaN_3) [31] results in formation of carbon nitride with increased ($C/N = 0.72$) and lowered ($C/N = 0.80$) contents of nitrogen accordingly.

Despite of so scale application of carbon nitride in various technologies intensive attempts recently are undertaken to expand and to improve parameters of its properties at the expense of introduction in graphite-like structure of atoms of non-metals, in particular, sulfur, phosphorus [32] or oxygen [29]. Let's remind, that the doping of molecules of fullerenes and carbon nanotubes with atoms of nitrogen, boron and silicon has ensured the creation of heterocarbon [33–43] and formation, in particular, at aza- and borafullerenes as well as at NC-nanotubes of extremely rare electronic properties. Doping of g-C_3N_4 by oxygen with formation O-C_3N_4 has allowed essentially to improve, for example, it fotocatalytic properties [29]. It is supposed [29], that in a lattice of O-doped carbon nitride there is an own modulation of electronic and zoned structure, therefore the border of photoactivity O-C_3N_4 in a visible spectrum extends up to 498 nm and more [29]. Two-stage synthesis of O-C_3N_4 with the contents of nitrogen close to formula (0.751) [29] is carried out by a hydrothermal method previously synthesized g-C_3N_4. At the first stage carbon nitride receive by heating of dicyandiamide at 550 °C within 5 h in a flow of argon. Then the synthesized powdery g-C_3N_4 disperse in 30 % a water solution of hydrogen peroxide also maintained within 10 h in teflon hermetic autoclave at 140 °C. Then a product was precipitated on centrifuge and washing with water. Obtained by given method sample of O-C_3N_4 contained (on the data XPS) 7.98 % O [29].

In the given report are presented and discussed results of synthesis of O-doped of carbon nitride with high (~11.1 %) contents of oxygen by means of one stage method of thermal transformation of pyridine molecules. Research purpose: to develop new and simple method of preparation of O-doped carbon nitride and to demonstrate the features of it synthesis and properties concerning to carbon nitride g-C_3N_4 and O-doped carbon-onions.

9.2 Materials and Methods

With the purpose of preparation of carbon nitride as a precursors of O-C_3N_4in quality of a nitrogen containing (17.1 mass% of nitrogen) reagent was used pyridine, which especially frequently is investigated in reactions of continuous flowing pyrolysis (CFP) for synthesis of the various forms of N-heterocarbon [44]. So, at pyrolysis of pyridine vapours were received N-graphene, N-carbon multiwalled nanotubes and fibres, the contents of nitrogen in which make 0.6, 2.0 and 2.5 mass% accordingly [44]. The spherical particles (by a diameter from 40 up to 1000 nm) of N-doping carbon containing up to 5 mass% of nitrogen by means of CFP of pyridine at 1000 °C were obtained [45]. The substances with the small contents of nitrogen such as acrylonitrile (C_3H_3N), benzonitrile (C_7H_5N), methane, hydrogen, acetylene and hydrogen cyanide were detected by a chromatographic method as the basic products at CFP of pyridine in an interval of temperatures 677–1127 °C [46]. Notably, that at products of pyrolysis of N-containing organics always hydrogen cyanide (HCN) presents [46]. Thus it is supposed [46], what the formation of N-heterocycles is initiated by pyridine radicals (C_5HN•), whereas the soot at the expense of reactions of olygomerization is formed. It is important to note, that attempts to receive carbon nitride by means of pyrolysis of molecules of pyridine have appeared in general unsuccessful [22].

Such low contents of nitrogen in products of CFP of pyridine is connected, first of all, that the formation and localization of heterocarbon is fulfilled only in the most high-temperature part of reactor, where are fulfilled basically endothermal heterogeneous (on walls of reactor) reactions of destruction of precursor molecules. The nitrogen as HCN formed at destruction of C_5NH_5 is leaved from this high-temperature zone of reactionary space, where are realized mainly heterophase (molecule, radicals – nanostructures) reactions of formation of carbon nanostructures, which, naturally, contain small quantity of nitrogen.

In an author's method of pyrolysis (AMP) [33–43] (distinct from CFP and flesh-pyrolysis (FP)) the products are investigated which are formed only in volume of reactor, are taken out by a reactionary flow from a high-temperature zone and are located in low-temperature (<70 °C) zone of reactionary space. As was established by us earlier [33–42] in reactionary conditions of AMP is realized not only process of fullerenization of pyridine molecules with formation of the condensed substances, in particular, as homoatomic (quasi-fullerenes (C_{48} and C_{40}) and fullerene (C_{60})) and heteroatomic, nitrogen containing (>12 % of nitrogen) closed spheroidal molecules (polyazafullerenes). Is remarkable, that from molecules of pyridine in AMP at ~1000 °C is formed also N-heterocarbon as N-doped onion carbon with high (~15 %) contents of nitrogen [33–36]. The main feature of AMP that condensed (molecules) and deposited (soot) the products are formed in a vapour-gaseous phase at the expense of radical reactions, in particular, polymerization and polycondensation of activated molecules of pyridine and fragments of their partial destruction.

For optimization of conditions of even greater (more than 15 % of nitrogen) nitrogenization of heterocarbon with the purpose of obtaining of carbon nitride the pyrolysis of pyridine vapours is carried out in a wide range of temperatures (950–1050 °C) at a variation of stay time of reagents in reactionary space. The experiences were carried out on typical for a method of continuous flowing pyrolysis of organics vapours set up with use tubular reactor [33–42]. Comparison of the products received at fixed reactionary conditions estimated by their quantity, contents of nitrogen and XRD analysis. For the further research samples were selected, XRD patterns which corresponded only to graphite-like carbon nitride. Results of research of a sample **A** as average from products eight experiences received at 1050 °C and located in a low-temperature zone of reactionary space here are submitted.

As earlier [33, 36] by us were obtained at pyrolysis also of pyridine samples **B** and **C** of N-doped carbon-onions with the contents of nitrogen 14.7 % and 10.5 % accordingly, it was important their properties to compare to a sample **A** of O-C_3N_4. Moreover, for a correctness of results of comparison of properties of new substances (samples **A**, **B** and **C**) was used as a test sample **D** of the carbon nitride synthesized by means of thermal treatment of melamine on a technique described in [18] for obtaining closed (C/N = 0.68) to formula (C/N = 0.75) composition g-C_3N_4. So, melamine heated up with speed 20 °C/min up to 500 °C, maintained at this temperature during 4 h, and then with speed 10 °C/min the temperatures 520 °C reached and maintained even 2 h.

Sample **A** after extraction from it soluble in water and benzene of substances and test sample **D** of g-C_3N_4 investigated by methods of chemical and XRD (Dron UM-1 with Cu K_α – radiation and nickel filter) analysis as well as IR spectroscopy. A sample A was investigated also by methods X-ray photoelectron spectroscopy and temperature-programmed desorption with the mass spectrometric analysis (TPD MS) of formed products. Mass spectrometric analysis of volatile products of thermolysis in range up to 200 m/z was carried out on exclusive mass spectrometer MX-7304A (Sumu, Ukraine) with ionization by electrons (EI) [47]. Programmed linear heating of a sample up to 750 °C carried out with speed 0.15 K s^{-1}. IR spectra in a regime of reflection were written down in a range from 4000 up to 400 sm^{-1} (the sanction is 8 sm^{-1}) with use of spectrometer Nicolet Nexus FTIR (Thermo Scientific). X-ray photoelectron spectra (XFS) of samples were measured on photoelectronic spectrometer of manufacture SPECS Surface Nano Analysis Company (Germany) with the half-spherical analyzer PHOIBOS.

9.3 Results and Discussion

9.3.1 *X-Ray Diffraction and Chemical Analysis of the O-Doped Carbon Nitride*

First of all, it is necessary to note, that diffractograms of samples **A** (Fig. 9.2, curve 2) and **D** (Fig. 9.2, curve 1) are enough well coordinated to an experimental X-ray spectrum for g-C_3N_4 as well as with calculated data [1, 3, 18, 48]. The sample **A** contains 23.7 mass% of nitrogen, 11.1 mass% of oxygen, 1.8 mass% hydrogen and 63.4 mass% of carbon, that corresponds to formula composition $C_3N_{0.96}O_{0.39}H_{1.02}$ with a ratio C/N = 3.13. The presence of such large quantity of oxygen at a sample **A** has appeared completely unexpected, as pyrolysis of pyridine vapours is carried out in oxygen-free atmosphere. Samples of carbon nitride received by different methods in oxygen-free atmosphere, usually, contain up to 2–3 mass% of oxygen, which, as it is accepted to consider, is chemisorbed from air. It is possible to assume, that formed by means of radical reactions in a vapour-gaseous flow heptazine layers from fragments C_6N_7 not only structurally are disorder because of a high gradient of temperatures (from 1050 up to 70 °C), but also contain valence-unsaturated atoms of carbon, which as radicals on air by oxygen are deactivated. Is remarkable, that received by us earlier from pyridine N-doped carbon-onions [33, 36] also contained up to 10 % of oxygen. At the same time, soot, which in parallel with a sample A is formed at pyrolysis of pyridine, but is deposited in the most high-temperature zone, contains no more than 2.4 % of nitrogen and no more than 1.9 % of oxygen. Let's note that the oxidation of samples of carbon nitride synthesized in stationary

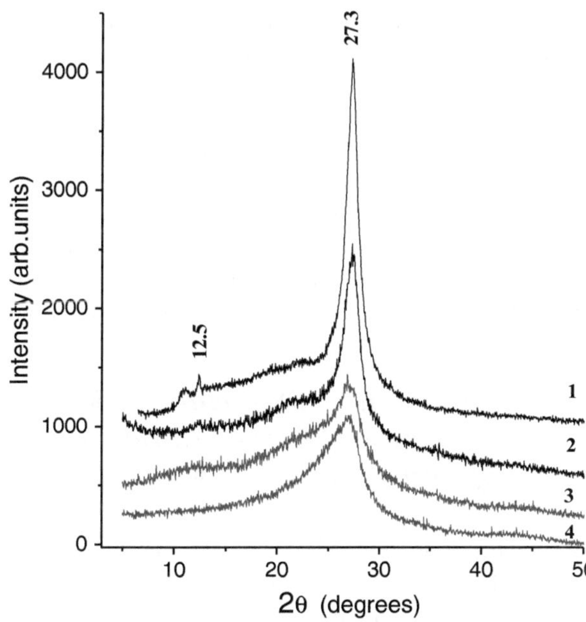

Fig. 9.2 X-ray diffraction of samples: 1 – **D**, 2 – **A**, 3 – **B**, 4 – **C**

conditions with the purpose of obtaining O-doped C_3N_4 is carried out in a solution of hydrogen peroxide [29].

In XRD patterns of samples **A** and **D** the most intensive line at $2\theta = 27.3°$ (is usually characteristic for graphite-like materials) is caused by interlayer stacking of aromatic rings. However interplanar distance between the nearest layers (d002) in a sample **A** it is a little less (0.326 nm), than at crystal graphite (0.353 nm), and, hence, in a sample $C_3N_{0.96}O_{0.39}H_{1.02}$ density of packing in perpendicular to a direction of layers is above. It is supposed [3, 18, 48], that it is connected with stronger van der Waals bonds in adjacent layers of g-C_3N_4. How it is possible to see (Fig. 9.2), intensity of a reflex ($2\theta = 27.3°$) in a sample **A** (curve 2) is much higher in comparison with samples N-doped carbon-onions (curves 3 and 4). Probably, it is connected to the rather smaller size crystallites in samples N-doped carbon. Some broadening of a basic diffraction line (002) in a spectrum of a sample **A** in comparison with a sample **D** (Fig. 9.2, curve 1) can display heterogeneity on the size nanocrystallites of a sample **A**.

On X-ray spectrum of a sample **A** also is present broad very weak peak at $2\theta = 12.5°$ (d001 = 0.707 nm), which shows periodicity of stacking of heptazine fragments in a plane d002. The value 0.707 nm is less some than counted size of one tri-*s*-triazine block (d001 = 0.713 nm), that can be caused by non-planarity of heptazine layers d002. Value 2θ of peak (001) on X-ray spectra of samples of carbon nitride usually is changed in an interval 12.6–13.2° (d001 = 0.699–0.670 nm) [1, 3, 18, 48] and also depends on a method of its synthesis. As the peak (001) is characteristic for X-ray spectrum of melones $(C_6N_9H_3)n$, constructed from heptazine fragments, its presence at a X-ray spectrum of carbon nitride **D** can testify, that the plane d002 also consists of basically heptazine (instead of triazine) rings [3]. The peak 001 in a sample **A** is shown very poorly because of the small contents in it of nitrogen (23.7 % N). In X-ray spectrum of N-doped carbon-onions (Fig. 9.2, curve 3), containing of nitrogen less than 15 %, the peak (001) at $2\theta \sim 12.5°$ practically is not shown, and in a spectrum of a sample **C** (~10 % N) (Fig. 9.2, curve 4) it completely is absent. Therefore it is logical to assume, that the plane d002 in a sample **A** can consist mainly not of heptazine fragments (C_6N_7), and from containing two atoms of nitrogen of pyrazine (C_4N_2, 1,4 -) and pyrimidine (C_4N_2, 1,3 -) cycles, which worsen planarity of a plane d002.

Thus, synthesized from pyridine the sample $C_3N_{0.96}O_{0.39}H_{1.02}$ is oxygen-doped carbon nitride (O-g-C_3N), as in its structure (structure of carbon nitride) pyrazine and pyrimidine the layers in addition are contained with atoms of oxygen.

9.3.2 Peculiarities of Structure of IR Spectra of Doped by Oxygen of Carbon Nitride (O-g-C_3N) and by Nitrogen of Carbon-Onions as well as Carbon Nitride

IR spectra (Fig. 9.3) of O-g-C_3N (curve 3) and g-C_3N_4 (curve 4) as against samples **B** and **C** of N-doped carbon-onions (curves 2 and 1) contain distinctly seen intensive absorption bands at 1642, 1599, 1498, 1351, 1252 cm^{-1}, which, usually,

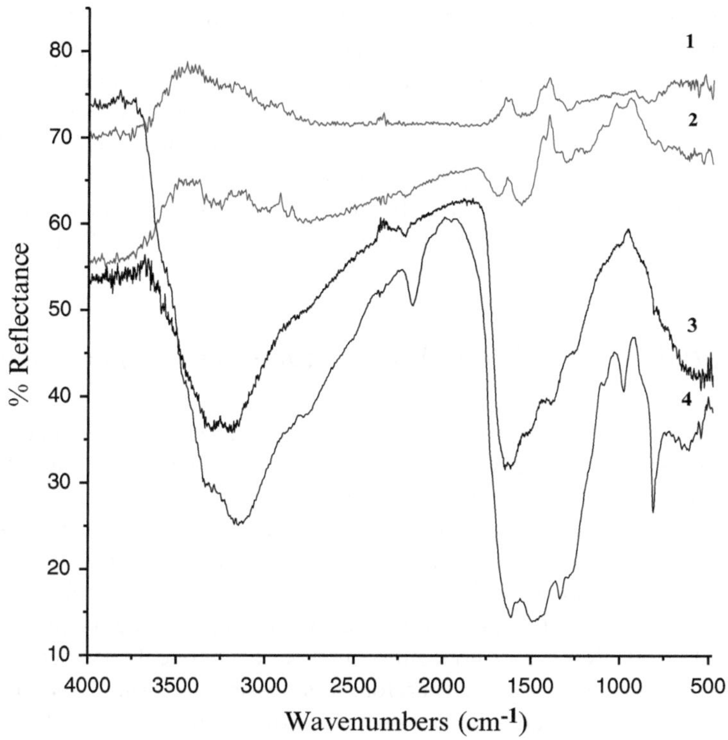

Fig. 9.3 IR spectra of spectra of the samples **C**, **B**, **A** and **D** (curves 1, 2, 3 and 4 respectively)

are characteristic for stretching vibrations of aromatic CN-bonds in condensed nitrogen containing heterocycles. So group of peaks in the range of 1200–1650 sm[1] correspond to CN-bonds not only in carbon nitride, but also in melem (2,6,10-triamino-*s*-heptazine $C_6N_7(NH_2)_3$), heptazine $((C_6N_7)\ H_3)$ and melon $(C_6N_9H_3)_n$ ([3, 18, 48]. In particular, absorption bands at 1336 and 1641 cm^{-1} can answer stretching vibrations of bonds C—N and C═N accordingly [48]. Is remarkable, that distinctly observable in IR spectrum of a sample **D** (Fig. 9.3, curve 4) absorption bands at 809 cm^{-1}, characteristic for triazine cycle $(C_6N_7)_n$ ([3, 18, 48], in IR spectra of O-g-C_3N (Fig. 9.3, curve 3) and samples **B** and **C** (Fig. 9.3, curves 1, 2) is shown as extremely weak peaks. The low intensity of this absorption band in IR spectrum of O-g-C_3N is caused by that the plane 002 in its structure according to X-ray spectrum (Fig. 9.2) consists of pyrazine and pyrimidine of layers and practically does not contain heptazine fragments (C_6N_7).

Small intensive peak at ~2220 cm^{-1} in IR spectra of O-g-C_3N and g-C_3N_4 is an evidence of presence of insignificant quantities of terminal cyano groups (—C≡N) in their structures [3]. The wide and intensive absorption band at 3306 cm^{-1} corresponds stretching vibrations of terminal amino groups (—NH_2), whereas the band at 3211 cm^{-1} can belong to stretching vibrations of —OH groups. It is important to note, that in weak structurizated IR spectra of samples **B** and **C** of N-doped carbon-onions (Fig. 9.3, curves 1 and 2), which contain smaller quantities of nitrogen and oxygen, the absorption bands of stretching vibrations of —NH and —OH bonds are shown extremely poorly. In the IR spectrum of O-g-C_3N there is also weak peak at 1080 cm^{-1} adequate C—O of bonds. Is remarkable, that characteristic for carbonyl group (C═O) the absorption bands at ~1700 cm^{-1} in the IR spectrum of O-g-C_3N is not obviously shown, as it as a weak band can be blocked (or to coincide) by extremely intensive peak of bond C═N at 1642 cm^{-1}.

9.3.3 Study of O-g-C_3N by X-Ray Photoelectron Spectroscopy

The study of the synthesized sample **A** by method of XPS has shown, that it consists only of carbon, nitrogen and oxygen (Fig. 9.4a) and contains a little bit smaller (rather data of the element chemical analysis) quantities of nitrogen (21.3 mass%) and oxygen (8.9 mass%). Really, in a XPS spectrum of O-g-C_3N (Fig. 9.4a) there are three different on intensity of peaks, characteristic for binding energy C1s (~285 eV), N1s (~400 eV) and O 1 s (532 eV). Is remarkable, that all three spectral lines (C 1 s, N 1 s and O 1 s) represent wide and asymmetric peaks, and, hence, in structure of O-g-C_3N there are nonequivalent types of chemical bond between atoms of carbon, nitrogen and oxygen. Therefore it is possible to assume, that peak C 1 s (Fig. 9.4b, curve 1) includes lines with values of binding energy ~284.7, 286.0, 288.3 and 289.4 eV, which, in view of the data [29], correspond to bonds C—C, C—NH_2, C—N and C—O accordingly. So, in [29] is shown, that in broadened peak C 1 s of O-doped g-C_3N_4 there are four lines with values of binding energy 284.6 (C—C), 286.1 (C—NH_2), 288.0 (C—N) and 289.0 eV (C—O). However

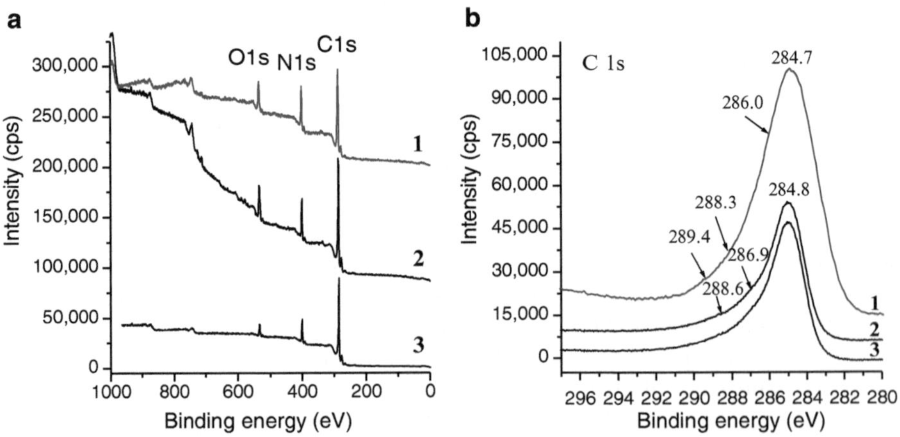

Fig. 9.4 XPS survey (**a**) and core level C 1 s (**b**) spectra of the samples **A**, **B** and **C** (curves 1, 2 and 3 respectively)

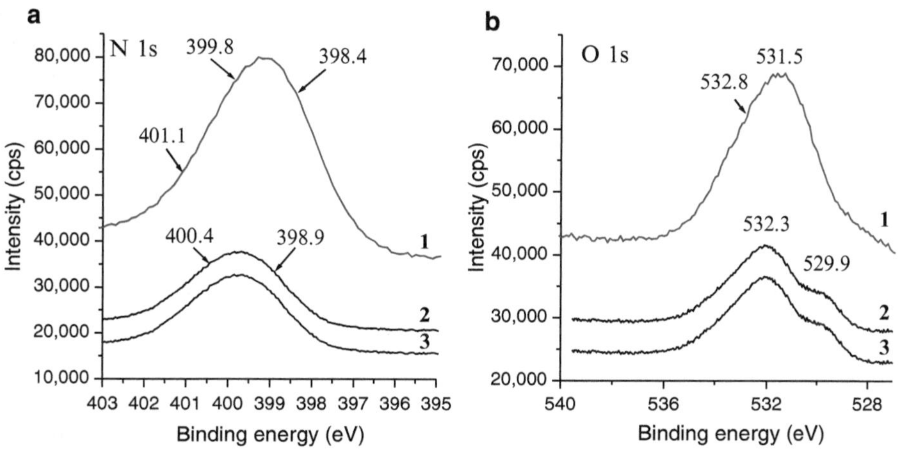

Fig. 9.5 XPS core level N 1 s (**a**), O 1 s (**b**) spectra of the samples **A**, **B** and **C** (curves 1, 2 and 3 respectively)

in a spectrum C 1 s of carbon nitride close to formula composition and, hence, consisting mainly from heptazine fragments C_6H_7, there are only two lines of carbon with binding energy 284.7–285.7 and 287.8–288.2 eV, which are responsible for bonds with two- and three-coordinated atoms of nitrogen accordingly [1, 3, 49]. Spectral C 1 s the line at 289.4 eV adequate to bond C—O practically is not shown (Fig. 9.5b, curves 2, 3) in spectra of N-doped carbon-onions [33, 36], containing on the data of XPS it is less both nitrogen, and oxygen (in a sample B: N – 16.8 mass%, O – 7.7 mass%, in a sample C: N – 13.6 mass%, O – 3.9 mass%).

In broad asymmetric peak N 1 s of XPS spectrum (Fig. 9.5a, curve 1) of a sample O-g-C_3N it is possible to see two strongest lines with values 398.4 and 399.8 eV binding energy. These values, probably, belong to accordingly two-coordinated (sp^2, pyridine) and three-coordinated (sp^3) atoms of nitrogen. The received results for O-g-C_3N are well coordinated to values 398.5 (sp^2) and 399.8 (sp^3) eV of binding energy N 1 s in O-doped carbon nitride (C/N = 0.751), containing 7.98 at.% of oxygen [29]. In a spectrum N 1 s of a sample O-g-C_3N it is possible also to see weak peak with binding energy 401.1 eV, which, as believes in ([29], is characteristic for nitrogen of amino group (C—NH_x) in O-doped carbon nitride. The values of binding energy describing bonds of carbon from two- and three-coordinated atoms of nitrogen, as well as with nitrogen of amino group, are present also in N 1 s spectrum of non-carbon nitride: 398.5–398.7 (C—N—C), 400.0–400.1 (N—C_3), 401.4 (C—NH_x) [1, 3, 49]. However in a spectrum N1s of N-doped carbon-onions [33, 36] to pyridine (399,1 eV) and quaternary (400.4 eV) atoms of nitrogen correspond to lines with some larger (on 0.6 eV) binding energy (Fig. 9.5a, curves 2 and 3). It is considered [3] that the similar decrease of values of binding energy N1s can be by the result of redistributions of additional electrons in heptazine layer of carbon nitride.

The asymmetry of peak O 1 s (Fig. 9.5b, the curve 1) in XPS spectrum of O-g-C_3N can be a consequence of a superposition of two lines. The line with binding energy 532.8 eV answers, as it is accepted to consider [3, 29, 49], oxygen of adsorbed molecules of water on a surface of sample. A line with binding energy 531.5 eV can answer C—O bond, as in a spectrum O 1 s of O-doped carbon nitride (O- 7.98 at. %) the binding energy N—C—O makes 531.6 eV [29].

9.3.4 Mass Spectrometric Study of Composition of Gaseous Products of Thermolysis of O-g-C_3N

Thermostability of carbon nitride and composition of products it thermolysis at various temperatures practically were not investigated yet, though these characteristics are extremely important for use of carbon nitride, in particular, as support and catalyst. In Pozdnjakov et al. [50] it is marked, that gaseous products of thermolysis of a powdery carbon nitride g-C_3N_4 (including, in opinion [50], as an impurity and α-C_3N_4 modification) are detected (by a mass spectrometric method) already at 500 °C, though its complete decomposition is finished at 850 °C. The thermolysis products of such sample of carbon nitride consist of mainly CN fragments, such $C_3N_2^+$ (m/z 64), $C_2N_3^+$ (m/z 66), CN_4^+ (m/z 68), $C_3N_3^+$ (m/z 78), $C_3N_4^+$ (m/z 92) [50].

Here it is necessary especially to note, what the authors [50] are not lonely in detecting by means of X-ray diffraction ostensibly of crystal phases of carbon nitride, in particular, superhard β-C_3N_4 with sp^3-bonded carbon. However such compounds of carbon with nitrogen from the chemical point of view cannot be formed basically, as have confirmed almost 20-year's numerous and intensive (with use of the most rigid (thermobaric) methods of influence on substance), but unsuccessful attempts of their obtaining. Theoretical counts which have stipulated for possible existence and synthesis of hypothetical super hard β-C_3N_4 were based only on such parameters as the size of nitrogen and carbon atoms and polarity of C-N bond that is not absolutely correct. First of all, according to such formalistic approach a compound of boron (boron atom is less than nitrogen atom) with carbon (or oxygen) should be even more superhard, than hypothetical β-C_3N_4. However reality is perfect another: boron carbide B_4C and boron suboxide B_6O [40] considerably concede on hardness not only diamond, but also cubic boron nitride (c-BN). Besides, a choice of silicon nitride β-Si_3N_4 as the prototype for account of hypothetical crystal carbon nitride is more than formal and unreasonable. According to similar logic silicon nitride should exist as SiN as analogue of silicon carbide (SiC). However Si_3N_4 is existed only.

And, at last, there is the most important argument, why crystal C_3N_4 cannot be synthesized basically. The atom of silicon as against atom of carbon is capable to locate (and locates, for example, in a plane configuration of trisilylamine $N(SiH_3)_3$) on empty extremely low energetic 3d-orbital not binding electrons of atom of nitrogen with formation additional (to three σ-bonds N—Si) π-bond N—Si. This additional interaction as unique multiple bond between atoms of nitrogen and silicon promotes the greater strengthening of N-Siσ-bond, therefore theoretically pyramidal skeleton NSi_3 in a molecule $N(SiH_3)_3$ gets a plane configuration. This case is not sole: the interaction of atom of silicon with oxygen and phosphorus in molecules of ($O(SiH_3)_2$) ($P(SiH_3)_3$) respectively is characterized also additional double pπ-dπ-bond. Therefore it is possible to assume, what in structure of β-Si_3N_4 multiple interaction N—Si also is realized, not binding electrons of atom of nitrogen are located on this bond and it is strengthened. The atom of carbon (as against atom of silicon) has not at all d-orbital. Therefore at formation σ-bond C—N not binding electrons of atom of nitrogen are delocalized and only weaken this bond that predetermines existence of a molecule of trimethylamine $N(CH_3)_3$ as analogue of a flat molecule $N(SiH_3)_3$ in a pyramidal state. Thereforeβ-C_3N_4 as against β-Si_3N_4 cannot exist because of absence of an opportunity by carbon atom to locate not binding electrons of nitrogen atom.

Mass spectrometric research which has been carried out in an interval of the temperatures 50–750 °C, has shown, that an initial product of decomposition of O-g-C_3N is HCN (Fig. 9.6), ion (m/z 27) which, is presented at products of thermolysis in all the investigated interval of temperatures. The destruction of a molecule HCN with formation of an ion CN^+(m/z 26) is observed with 200 up to 750 °C. Registered at temperatures lower 150 °C the ions with m/z 18, 28 and 44 (Fig. 9.6) can correspond to desorbed H_2O, CO and CO_2 molecules. The basic products of

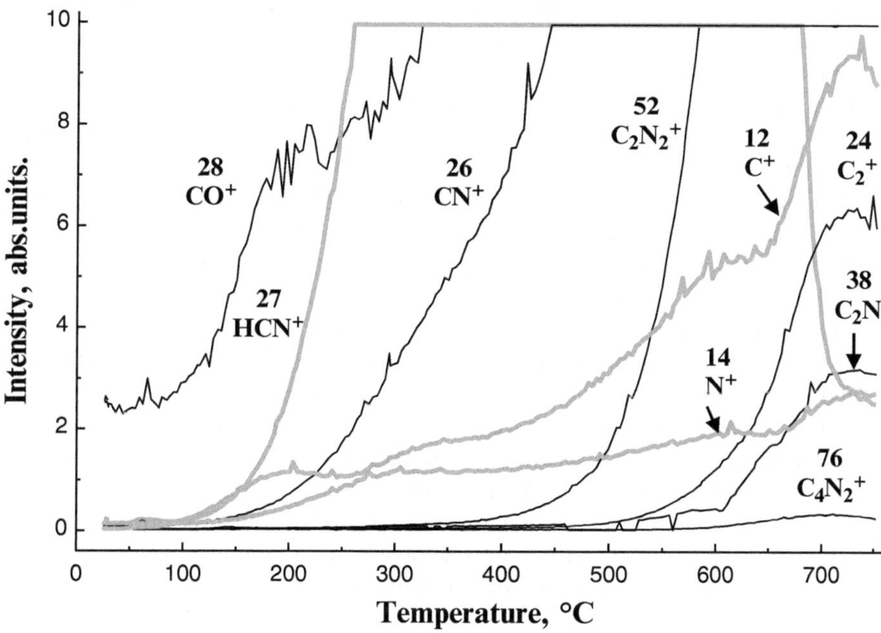

Fig. 9.6 Experimental thermodesorption curves for selected 12, 14, 24, 26, 27, 28, 38, 52 and 76 amu peaks for the sample **A**

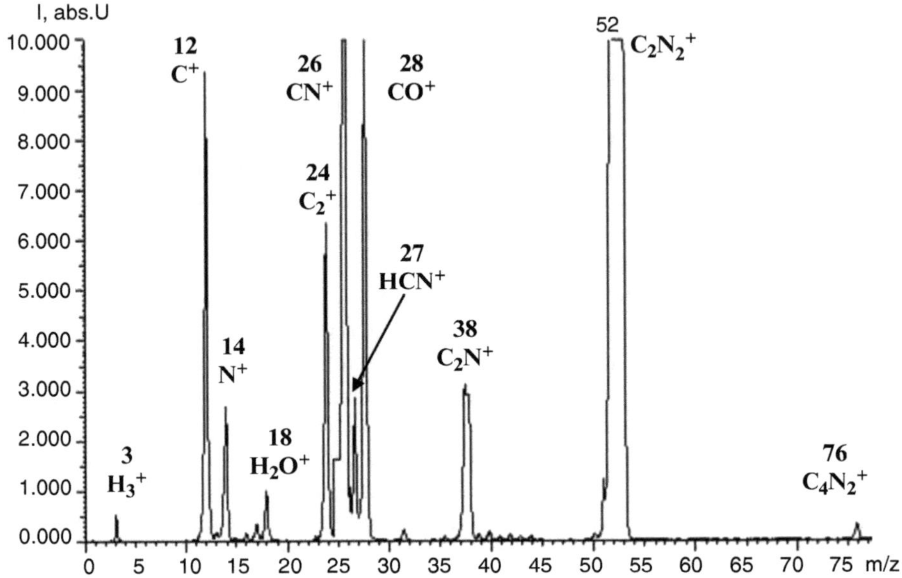

Fig. 9.7 Mass spectrum EI of the products of the thermolysis of the sample **A** at the temperature 725 °C

thermolysis O-g-C_3N down to 400 °C are HCN, CN^+ and CO (рис.6), which can be as the appropriate functional groups (C=O, C≡N, C=NH) are located on defects of pyrazine and pyrimidine cycles.

At temperatures of thermolysis of O-g-C_3N higher 500 °C are observed the destruction of pyrazine and pyrimidine cycles with allocation mainly cyanogen, which ion $C_2N_2^+$ (m/z 52) in mass spectra (Fig. 9.7) is characterized by the greatest intensity. At higher temperatures (600–750 °C) in mass spectra are detected (Fig. 9.7) alongside with equiatomic ion $C_2N_2^+$ ions $C_4N_2^+$ (m/z 76) and C_2N^+ (m/z 38). Is remarkable, that in products of thermolysis the ions C_2^+, C^+ and N^+, which peaks in mass spectra (Fig. 9.6) rather intensive that are detected also. Hence, at temperatures higher 600 °C at destruction of nitrogen – carbon fragments the atoms of nitrogen and carbon are formed (It is possible, certainly that the ions $C^+$and N^+ are generated under action of ionization electronic impact).

Thus, O-doped carbon nitride O-g-C_3N as against carbon nitride g-C_3N_4 is less thermostable (probably because of it nanodispersity). Thermolysis of a sample O-g-C_3N is carried out with formation mainly cyanogen and fragments with the large contents of atoms of carbon ($C_4N_2^+$and C_2N^+), that is quite logical at decomposition of a product with mainly pyrazine and pyrimidine cycles containing two (instead of three as in g-C_3N_4) of atoms of nitrogen.

9.4 Conclusions

1. The new and simple method of synthesis of a nanodispersed powder of O-doped carbon nitride (O-g-C_3N) with high (~11.1 %) contents of oxygen from pyridine is developed.
2. Carbon nitride at pyrolysis of pyridine vapours is formed only in reactionary gaseous flow, is taken out by this flow from high-temperature (1050 °C) zone of reaction and is located in low-temperature (<70 °C) zone of reactionary space.
3. Resultant carbon nitride structurally is disorder because of a high gradient of change of temperature (with 1050 up to 70 °C) and contains in pyrazine and pyrimidine cycles of valence-unsaturated atoms, which of air easily join atoms of oxygen that promotes formation O-g-C_3N. Valence-unsaturated carbon nitride is formed mainly of molecules HCN (or CN-radicals), which are generated at thermal destruction of molecules C_5H_5N.
4. The thermal decomposition of O-g-C_3N is began at temperatures above 500 °C. In mass spectra of products it thermolysis are detected not only molecules (HCN, C_2N_2, and CO) and nitrogen – carbon fragments (CN^+, $C_4N_2{}^+$ and C_2N^+), but also atoms of carbon and nitrogen.

References

1. Thomas A, Fischer A, Goettmann F (2008) Graphitic carbon nitride materials: variation of structure and morphology and their use as metal-free catalysts. J Mater Chem 18:4893–4908
2. Zhang S, Li J, Zeng M et al (2014) Polymer nanodots of graphitic carbon nitride as effective fluorescent probes for the detection of Fe^{3+} and Cu^{2+} ions. Nanoscale 6:4157
3. Dai H, Gao X, Liu E et al (2013) Synthesis and characterization of graphitic carbon nitride sub-microspheres using microwave method under mild condition. Diamond Relat Mater 38:109–117
4. Zheng Y, Liu J, Liang J et al (2012) Graphitic carbon nitride materials: controllable synthesis and applications in fuel cells and photocatalysis. Energy Environ Sci 5:6717–6731
5. Cao SW, Liu XF, Yuan YP et al (2013) Artificial photosynthetic hydrogen evolution over g-C 3 N 4 nanosheets coupled with cobaloxime. Phys Chem Chem Phys 15:18363–18366
6. Martha S, Nashima A, Parida KM (2013) Facile synthesis of highly active g-C3N4 for efficient hydrogen production under visible light. J Mater Chem A 1:7816–7824
7. Zhang Y, Pan Q, Chai G et al (2013) Synthesis and luminescence mechanism of multicolor-emitting g-C_3N_4 nanopowders by low temperature thermal condensation of melamine. Sci Rep 3(1943):1–8
8. Han J, Xu G, Ding B et al (2014) Porous nitrogen-doped hollow carbon spheres derived from polyaniline for high performance supercapacitors. J Mater Chem A 2:5352–5357
9. Widlow I, Chung YW (2000) Recent progress in synthesis and characterization of amorphous and crystalline carbon nitride coatings. Braz J Phys 30(3):490–498
10. Gao D, Xu Q, Zhang J et al (2014) Defect-related ferromagnetism in ultrathin metal-free g-C3N4 nanosheets. Nanoscale 6:2577–2581
11. Zhang X, Zhao M, Wang A et al (2013) Spin-polarization and ferromagnetism of graphitic carbon nitride materials. J Mater Chem C 1:6265–6272
12. Demichelis F, Rong XF, Schreiter S et al (1995) Deposition and characterization of amorphous carbon nitride thin films. Diamond Relat Mater 4(4):361–365

13. Korsunskii BL, Pepekin VI (1997) On the way to carbon nitride. Russ Chem Rev 66(11):901–911
14. Fischer A (2008) Reactive hard templating. From carbon nitrides to metal nitrides. Ph.D. thesis, Potsdam
15. Majumdar A, Scholz G, Hippler R (2009) Structural characterization of amorphous hydrogenated-carbon nitride (aH-CNx) film deposited by CH_4/N_2 dielectric barrier discharge plasma: 13C, 1H solid state NMR, FTIR and elemental analysis. Surf Coat Technol 203:2013–2016
16. Franklin EC (1922) The ammono carbonic acids. J Am Chem Soc 44(3):486–509
17. Gu Y, Chen L, Shi L et al (2003) Synthesis of C_3N_4 and graphite by reacting cyanuric chloride with calcium cyanamide. Carbon 41(13):2674–2676
18. Katsumata K, Motoyoshi R, Matsushita N, Okada K (2013) Preparation of graphitic carbon nitride (g-C3N4)/WO3 composites and enhanced visible-light-driven photodegradation of acetaldehyde gas. J Hazard Mater 260:475–482
19. Zimmerman JL, Williams R, Khabashesku VN, Margrave JL (2001) Synthesis of spherical carbon nitride nanostructures. Nano Lett 1:731–734
20. Lu XF, Gai LG, Cui DL et al (2007) Synthesis and characterization of graphitic carbon nitride sub-microspheres using microwave method under mild condition. Mater Lett 61:4255–4258
21. Maya L, Cole DR, Hageman EW (1991) On the way to carbon nitride. J Am Ceram Soc 74(7):1686–1688
22. Bailey WH (2008) Graphite-like C_5N: synthesis, characterization, intercalation, and comparison with graphite analogs. Ph.D. thesis, Lehigh University, Bethlehem
23. Talapaneni SN, Mane GP, Mano A et al (2012) Synthesis of nitrogen-rich mesoporous carbon nitride with tunable pores, band gaps and nitrogen content from a single aminoguanidine precursor. Chem Sus Chem 5(4):700–708
24. Kouvetakis JJ, Sasaki T, Shen C et al (1989) Novel aspects of graphite intercalation by fluorine and fluorides and new B/C, C/N and B/C/N materials based on the graphite network. Synth Met 34:1–7
25. Kouvetakis J, Bandari A, Todd M et al (1994) Novel synthetic routes to carbon-nitrogen thin films. Chem Mater 6(6):811–814
26. Kouvetakis J, Kaner RB, Sattler ML et al (1986) A novel graphite-like material of composition BC3, and nitrogen-carbon graphites. J Chem Soc, Chem Commun 24:1758–1759
27. Li Y, Zhang J, Wang Q et al (2010) Nitrogen-rich carbon nitride hollow vessels: synthesis, characterization, and their properties. J Phys Chem B 114:9429–9434
28. Montigaud H, Tanguy B, Demazeau G et al (2000) C_3N_4: dream or reality? Solvothermal synthesis as macroscopic samples of the C_3N_4 graphitic form. J Mater Sci 35(10):2547–2552
29. Li J, Shen B, Hong Z et al (2012) A facile approach to synthesize novel oxygen-doped g-C3N4 with superior visible-light photoreactivity. Chem Commun 48:12017–12019
30. Guo Q, Xie Y, Wang XM et al (2003) Characterization of well-crystallized graphitic carbon nitride nanocrystallites via a benzene-thermal route at low temperatures. Chem Phys Lett 380:84–87
31. Guo Q, Xie Y, Wang XM et al (2004) Synthesis of carbon nitride nanotubes with the C(3)N(4) stoichiometry via a benzene-thermal process at low temperatures. Chem Commun 1: 26–27
32. Ma X, Lv Y, Xu J et al (2012) Strategy of enhancing the photoactivity of g-C_3N_4 via doping of nonmetal elements: a first-principles study. J Phys Chem C 116(44):23485–23493
33. Kharlamov A, Bondarenko M, Kharlamova G et al (2013) A new method of synthesis carbon with onion-like structure with high (10–13%) content of nitrogen from pyridine. Univ J Mater Sci 1(2):78–86
34. Kharlamov A, Bondarenko M, Kharlamova G (2013) Mass spectrometric research of hydrogenated molecules of carbon as products of pyrolysis of benzene and pyridine vapours. Chem Mater Eng 1(4):122–131
35. Kharlamov AI, Bondarenko ME, Kirillova NV (2012) New method for synthesis of fullerenes and fullerene hydrides from benzene. Russ J Appl Chem 85(2):233–238

36. Kharlamov AI, Kharlamov GA, Bondarenko ME (2013) Preparation of onion-like carbon with high nitrogen content (~15%) from Pyridine. Russ J Appl Chem 86(10):1493–1503
37. Kharlamov AI, Kharlamova GA, Bondarenko ME (2013) New products of a new method for pyrolysis of pyridine. Russ J Appl Chem 86(2):167–175
38. Kharlamov AI, Kharlamova GA, Bondarenko ME (2013) New low-temperature method for joint synthesis of C_{60} fullerene and new carbon molecules in the form of C_3-C_{15} and quasi-fullerenes C_{48}, C_{42}, C_{40}. Russ J Appl Chem 86(8):1174–1183
39. Kharlamov O, Bondarenko M, Kharlamova G et al (2015) Nanoecological security of foodstuffs and human. In: Bonča J, Kruchinin S (eds) Nanotechnology in the security systems, NATO science for peace and security series C: environmental security, chapter 19. Springer, Netherlands, pp 215–229
40. Kharlamov AI, Kirillova NV, Loichenko SV, Kostyuk BD (2002) Properties of boron suboxide $B_{13}O_2$. Powder Metall Met Ceram 41:97–106
41. Kharlamova G, Kharlamov O, Bondarenko M (2013) In: Vaseashta A, Khudaverdyan S (eds) Advanced sensors for safety and security, NATO science for peace and security series B: physics and biophysics, chapter 31. Springer, Netherlands, pp 339–357
42. Kharlamova G, Kharlamov O, Bondarenko M, Fomenko V (2013) In: Vaseashta A, Khudaverdyan S (eds) Advanced sensors for safety and security, NATO science for peace and security series B: physics and biophysics, chapter 30. Springer, Netherlands, pp 329–338
43. Kharlamova G, Kharlamov O, Bondarenko M (2015) In: Bonča J, Kruchinin S (eds) Nanotechnology in the security systems, NATO science for peace and security series C: environmental security, chapter 20. Springer, Netherlands, pp 231–242
44. Teddy J (2009) CVD synthesis of carbon nanostructures and their applications as supports in catalysis. Ph.D. thesis, Toulouse University, Toulouse
45. Nikiwe K (2009) The synthesis of nitrogen doped carbon spheres and polythiophene/carbon sphere composites. Ph.D. thesis, University of the Witwatersrand, Johannesburg
46. Ninomiya Y, Dong Z, Suzuki Y, Koketsu J (2000) Theoretical study on the thermal decomposition of pyridine. Fuel 79(2):449–457
47. Kulik TV, Barvinchenko VN, Palyanitsa BB et al (2007) A desorption mass spectrometry study of the interaction of cinnamic acid with a silica surface. Russ J Phys Chem A 81(1):83–90
48. Xin G, Meng Y (2013) Synthesized g-C_3N_4 for photocatalytic degradation of methylene blue. J Chem 2013(1–5):5
49. Kim M, Hwang S, Yu J (2007) Novel ordered nanoporous graphitic C_3N_4 as a support for Pt–Ru anode catalyst in direct methanol fuel cell. J Mater Chem 17:1656–1659
50. Pozdnjakov RP, Blinov LN, Arif M (2005) Mass spectrometric study of carbon nitride C_3N_4. J Appl Phys 31(23):17–23

Chapter 10
Electrochemical Detection of 2,4,6-Trinitrotoluene at Colloidal Gold Nanoparticle Film Assemblies

Christopher P. Gulka, Evan A. Gizzie, David E. Cliffel, and David W. Wright

Abstract This work investigates citrate-capped, colloidal gold nanoparticle (AuNP) film assemblies of varying particle sizes (5–50 nm) adsorbed to bulk Au substrates to serve as platform electrochemical sensors to simultaneously detect and reduce TNT to 2,4,6-triaminotoluene (TAT) in solution. The high surface area-to-volume ratio of colloidal AuNPs offers advantages in electrocatalysis and enhanced signal transduction, while the facile immobilization onto a variety of substrates provides an adaptable and reproducible platform technology. In order to validate these AuNP film assemblies as platform sensors, square wave voltammetry is performed to enhance TNT sensitivity and optimize the signal transduction of TNT reduction. The highest sensitivity of TNT reduction is observed on 15 nm AuNP films (high nanomolar limits of detection). It is hypothesized that the 15 nm AuNP film assemblies store charge more efficiently because of their large dielectric constant compared to AuNP films of alternative sizes. It is also observed that upon assembly of the 5, 15, and 30 nm AuNPs, TNT reduction begins at more positive potentials compared to bulk Au and organic monolayer films, suggesting that these films exhibit electrocatalytic properties. The onset reduction potentials for 5, 15, and 30 nm AuNP assemblies were calculated and undergo 70 mV shifts to more positive potentials in comparison to bulk Au and organic monolayer control electrodes.

10.1 Introduction

Since World War I, 2,4,6-trinitrotoluene (TNT) has been the most widely utilized explosive, necessitating rapid and sensitive chemical detection [1]. The development of novel sensors for detection and identification of explosives is essential for improved security in the transportation industry, the neutralization of improvised explosive devices in current and former combat zones, and the elimination of

C.P. Gulka • E.A. Gizzie • D.E. Cliffel • D.W. Wright (✉)
Department of Chemistry, Vanderbilt University, Station B 351822, Nashville, TN 37235, USA
e-mail: David.wright@vanderbilt.edu

T.A. Camesano (ed.), *Nanotechnology to Aid Chemical and Biological Defense*,
NATO Science for Peace and Security Series A: Chemistry and Biology,
DOI 10.1007/978-94-017-7218-1_10

waste sites following the decommissioning of munitions factories. It is important to consider cheap materials entailing facile synthetic procedures for sensor development in order to more effectively streamline operation in developing regions where resources are scarce. Many existing sensors, microfluidics, and lab-on-a-chip devices are developed with electrochemical detection readouts because of the rapid response times for analysis and the sensitive detection strategies. Most traditional electrochemical schemes employ a three-electrode system comprising of a reference, counter, and working electrode [2]. While the reference electrode standardizes known potentials, the counter electrode acts as a cathode to assist in current flow between the working electrode and the buffer system. The reduction potential of the counter electrode also conveniently falls outside of the range of most chemical reactions. Most importantly, the working electrode bridges the gap between electricity and chemistry in order to probe reaction kinetics and thermodynamics at a conductive solid-water interface. Signal transduction is generated as a current output and is measured between the working and counter electrodes. However, innovative materials and approaches are essential to enhance signal transduction capabilities in sensors, so that detection schemes can be universally applied.

Gold has always garnered interest as a working electrode substrate because of its inert nature and facile molecular self-assembly [3–6]. In addition, gold conducts electrons efficiently, contains a large anodic potential range, can be manufactured easily, and requires little material for use [7]. In terms of platform sensor development, it is essential for electrons to be effectively shuttled between the electrode-buffer interfaces for optimum signal transduction, but one of the primary drawbacks of gold electrodes is the high background signal [8]. To alleviate these issues, organothiolate self-assembled monolayers (SAMs) have been utilized as insulators on gold electrodes to mask the high non-faradaic current to study the redox activity of the metalloproteins, cytochrome c and azurin [9–17]. Although adequate signal is provided to study the redox properties of these proteins, the insulating properties of organic thiols often compromises electron transfer kinetics and the signal is reduced. Leopold and coworkers have addressed these issues by adsorbing small monolayer-protected AuNP clusters onto SAMs in order to enhance the electron transfer kinetics, from enzyme to electrode, of cytochrome c and azurin redox proteins [18, 19]. The high-dielectric AuNP film assemblies provide a conductive substrate to perform electrochemical reactions, but also minimize background current in comparison to an unmodified gold substrate. It is hypothesized that electrons slowly tunnel through the low-dielectric (i.e. insulating) organic SAMs, while electrons rapidly hop from nanoparticle to substrate in the AuNP film assemblies, so that the signal is transduced rapidly [19]. Leopold further investigated the electrochemical properties of AuNP film assemblies by employing citrate-capped, colloidal particles as a hydrogen peroxide sensing platform [20]. The large volume AuNP increases the sensitivity of the platform, while the dense monolayer reduces non-Faradaic background. In addition, the simple and rapid film assembly procedure takes advantageous of stable, aqueous materials to offer low synthesis costs.

This chapter exploits the electrochemical and physical properties of electrostatically-bound, colloidal AuNP film assemblies in order to detect and electrocatalytically reduce TNT using voltammetric techniques. The explosive, TNT, possesses three known electrochemical reduction peaks, each of which is representative of the $6e^{-}/6H^{+}$ reduction process from a nitro to an amine group [21]. The energy of surface electrons in AuNPs is significantly higher in comparison to bulk gold, so that it is hypothesized that these films can exhibit electrocatalytic properties with smaller particles, thus increasing the reduction potentials of TNT and lowering the energy required to reduce the analyte in solution. Additionally, the sensing platform developed in this chapter only requires femtomoles of Au material, offering rapid and affordable synthetic approaches for sensor design and manufacturing.

10.2 Materials and Methods

10.2.1 Reagents

2,4,6-trinitrotoluene (TNT) was purchased from Cerilliant Corporation at a stock concentration of 1,000 ppm dissolved in acetonitrile. Citrate-stabilized AuNPs (15, 30, 50 nm) were purchased from Ted Pella, Inc. 5 nm citrate-stabilized AuNPs were purchased from NANOCS. Poly-L-Lysine (PLL) and 11-Mercaptoundecanoic acid (MUA) were purchased from Sigma-Aldrich. All other reagents were purchased from either Sigma-Aldrich or Fisher Scientific and used without further modification. All deionized water used in this investigation was filtered through a 0.22 μm filter at a resistance of 18.2 MΩ • cm.

10.2.2 Instrumentation

Electrochemistry was performed on a CH Instruments potentiostat with CHI660A analysis software. Surface UV-vis spectra were collected on a Varian Cary 5000 UV-vis-NIR spectrophotometer. Liquid-phase UV-vis measurements were performed on an Agilent 8453 spectrophotometer with a photodiode array detector. In order to probe surface topography, a Veeco Dektak 150 profilometer equipped with a diamond-tipped stylus was utilized. Images of AuNPs were captured with a Tecnai Osiris transmission electron microscope. Images of suspended AuNPs were captured with an iPhone 4S camera.

10.2.3 AuNP Film Assembly

Unmodified gold substrates evaporated on silicon wafers were first Piranha-cleaned (3:1 v/v ratio of concentrated H_2SO_4: 30 % H_2O_2) for 30 min and washed thoroughly with water. Next, a 5 mM MUA solution in ethanol was assembled for

2 days on the gold substrates. After washing with ethanol and water, the MUA-functionalized gold was immersed in a 1 mg/mL poly-L-lysine (PLL) solution for 20 min. After washing with water, the gold films were immersed in stock solutions of citrate-stabilized AuNPs (15 nm, 30 nm, 50 nm) for 1 h and subsequently washed with water. They were further stored in aqueous solutions for preserved stability. For the assembly of 5 nm AuNPs, the particles were centrifuged for 99 min at 16,100 × g once, and resuspended in 10 mM Tris buffer (pH 8.0). This particle solution was then drop-cast onto the MUA-PLL/Au substrate and allowed to dry overnight. The subsequent morning, the substrates were washed thoroughly with water.

10.2.4 Electrochemical Cell

The AuNP film immobilized on a Au substrate served as the working electrode ($A = 0.2\ cm^2$), a Pt wire served as the counter electrode, and a Ag/AgCl (3 M KCl) reference electrode comprised the electrochemical cell. The buffer was comprised of 50 mM phosphate buffer with 100 mM KCl (pH 7.0).

10.2.5 Double Layer Capacitance Measurements

Using the electrochemical cell described above, double-layer capacitance measurements were performed using cyclic voltammetry, scanning from +0.1 and +0.4 V vs. Ag/AgCl (beginning at +0.1 V) at a scan rate of 0.1 V/s in the aforementioned phosphate buffer.

10.2.6 Detection of 2,4,6-Trinitrotoluene at AuNP Film Assemblies

Electrochemical detection of TNT was performed using square wave voltammetry, scanning cathodically from −0.2 to −0.9 V vs. Ag/AgCl at a step potential of 4 mV, an amplitude of 25 mV, and a frequency of 15 Hz. TNT was dissolved in phosphate buffer containing 0.1 M KCl and 2.2 % v/v acetonitrile (pH = 7.0). Background measurements consisted of phosphate buffer with 0.1 M KCl containing 2.2 % v/v acetonitrile. TNT measurements were background-subtracted for optimized sensitivity.

10.3 Results and Discussion

10.3.1 Stability and Monodispersity of Colloidal AuNPs

The incorporation of citrate-stabilized AuNPs adsorbed to a poly-L-lysine (PLL) polymer provides a facile and rapid electrode assembly process exploiting electrostatic interactions using low-cost materials. Leopold et al. have assembled AuNP films on Au electrodes for enhanced electron transfer kinetics, but the size-dependent electrocatalytic activity of these assemblies has never been investigated [18–20]. In order to optimize the AuNP film for both sensing and catalysis, four different AuNP sizes are examined between 5 and 50 nm. There is a direct correlation between nanoparticle diameter and the ability to absorb light, as the molar absorptivity is directly proportional to the molecule or particle surface area. However, smaller particles contain a larger percentage of surface electrons, which possess a proclivity for catalysis. When suspended in aqueous buffers, the citrate particles demonstrate stability and monodispersity, as indicated by the captured TEM images, narrow plasmon resonance band in the UV-vis spectra, and bright red colors (Fig. 10.1). These aqueous, monodisperse particles serve as ideal building blocks for rapid and reproducible sensor development.

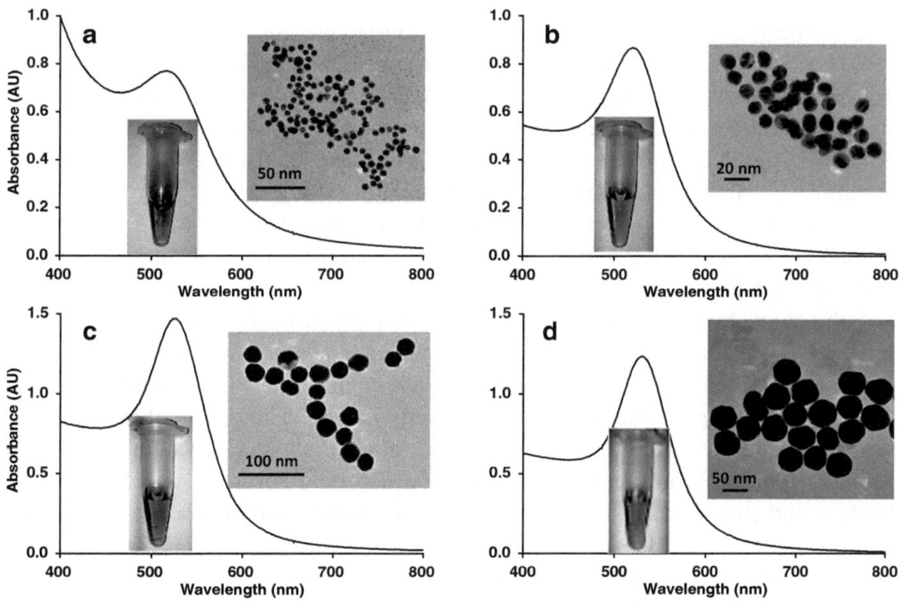

Fig. 10.1 Characterization of colloidal suspensions of citrate-capped AuNPs. UV-vis spectra, TEM images, and photographs of suspended AuNPs of (**a**) 5 nm, (**b**) 15 nm, (**c**) 30 nm, and (**d**) 50 nm AuNPs

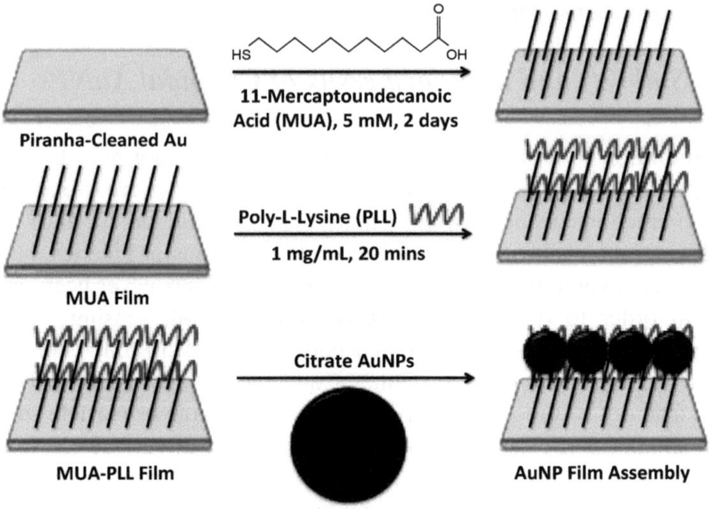

Fig. 10.2 Schematic of the assembly of citrate-capped AuNP films on Au substrates

10.3.2 AuNP Film Assembly and Characterization

The synthesis of the film is carried out according to previously described procedure [20]. Each gold substrate is first exposed to a fresh solution of Piranha for 30 min, which oxidizes all of the organic molecules adsorbed on the surface, increasing their water solubility and desorbing them from the surface. Next, a mercaptoundecanoic acid (MUA) alkanethiol is incubated with the clean gold for 2 days in order to assemble a stable monolayer on the surface (Fig. 10.2). Afterwards, a positively charged poly-L-lysine polymer (1 mg/mL) is incubated with the MUA/Au film for 20 min before incubating a stock solution of negatively terminated citrate-stabilized colloidal AuNPs for 1 h.

Film characterization is first carried out electrochemically by treating the AuNP assembly like an electrical-double layer capacitor. This can be performed using cyclic voltammetry to measure the background current (i.e. non-Faradaic current) at a potential range where little redox activity occurs. The potential range is first scanned from +0.1 to +0.4 V vs. Ag/AgCl at each electrode system (Fig. 10.3). First, it is interesting to note that an unmodified gold substrate possesses a high background current with significant variability, as indicated by the large error bars. With no material to mask natural interferents at the electrode surface, analytes such as water and oxygen can easily diffuse to the electrode, which attribute to the high background current. When organic insulating molecules such as MUA-PLL are adsorbed to gold, the capacitance decreases significantly with little variability, as indicated by the small error bars and the small total current measurements. The SAM creates a dense molecular blocking layer, shielding background effects of interfering analytes. Upon adsorption of the AuNPs to the MUA-PLL/Au substrates, the total

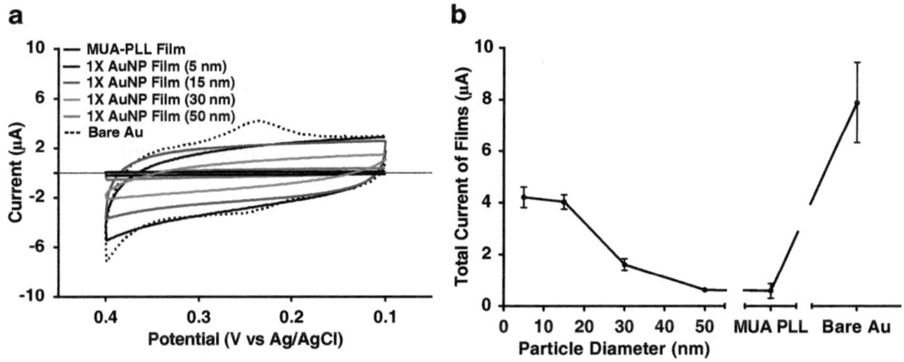

Fig. 10.3 Electrochemical capacitance measurements of various AuNP film assemblies. (**a**) Representative cyclic voltammograms of AuNP film assemblies, MUA-PLL film, and a bare Au electrode from +0.1 to +0.4 V vs. Ag/AgCl at a scan rate of 0.1 V/s and an electrode area of 0.2 cm^2. (**b**) Total current of each film or current thickness from the cyclic voltammograms obtained for each of the electrode assemblies performed in triplicate

current (i.e. current thickness) increases inversely proportional relative to particle diameter. The 50 nm AuNP film assemblies display identical total current values relative to the organic MUA-PLL. However, as particle size decreases, the total current increases. The 15 and 5 nm AuNP films exhibit similar levels of background current, which suggests that there may be a maximum threshold for non-Faradaic current for these colloidal film assemblies. Nonetheless, the AuNP films (5, 15, and 30 nm) exhibit an ability to store charge in a larger capacity relative to films of MUA-PLL and 50 nm AuNPs, while also simultaneously masking the surface from interfering analytes such as water and oxygen.

In order to further probe the AuNP assembly characteristics of this platform sensor, surface UV-vis spectra are generated to understand the particle behavior at the gold surface (Fig. 10.4). The 5, 15, and 30 nm AuNP films exhibit a localized SPR peak near 500 nm, which correlates to their SPR peak upon suspension in an aqueous solution. This indicates that not only are the particles adsorbed to the film, but they also maintain their monodisperse, catalytically active nature. In contrast, the 50 nm AuNP film assemblies possess two peaks: one at 500 nm and an additional at 625 nm. The presence of the additional peak indicates that the particles may undergo some level of aggregation at the electrode surface. The small current thickness and weak overall capacitance of the 50 nm AuNP film assembly is similar to that of an organic SAM, which suggests there is a threshold particle size for optimum packing and charge storing capabilities. Additionally, film thickness measurements of the assemblies via stylus profilometry reveal that the exploitation of the non-specific electrostatic adsorption of AuNPs does not create a single monolayer of particles at the electrode surface. For 5, 15, 30, and 50 nm AuNPs assemblies, the film thickness values are: 80, 100, 150, and 150 nm, respectively. This suggests that during the electrostatic adsorption process, the particles are non-specifically stacking on top of each other, which creates a thicker film than originally intended and a long-range electron transfer material.

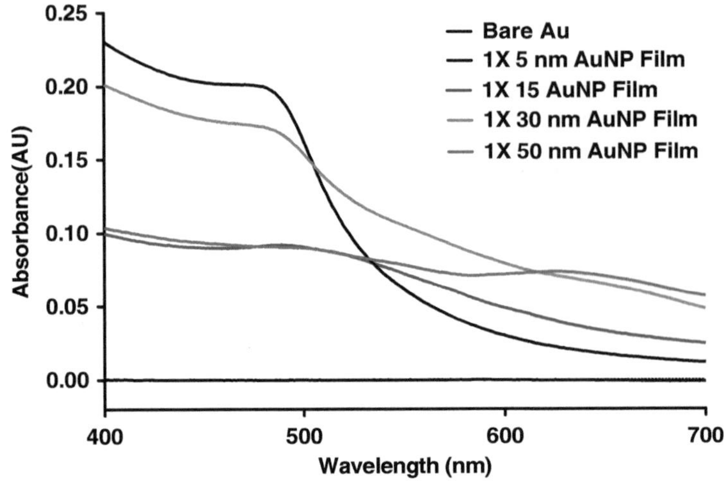

Fig. 10.4 Surface UV-vis spectra of AuNP film assemblies

10.3.3 Electrochemical Detection of TNT Using Platform AuNP Film Assemblies

Previous electrochemical strategies have developed complex, expensive electrodes for TNT detection, including but not limited to: boron-doped diamond, graphene supports, and innovative carbon-based supports [22–25]. Bulk gold substrates in addition to nanoparticles have been heavily investigated since the 1980s, as this accessible material provides an excellent sensing platform with commercially available materials at low purchasing costs. In order to determine the performance of these films, electrochemical TNT reduction is probed at each film assembly. The electrochemical technique, square wave voltammetry (SWV), is investigated because it provides better sensitivity than alternative voltammetry and polarography techniques. This pulsed voltammetry technique generates coupled forward and reverse steps along a staircase potential sweep, generating a single net current value for each scanned potential value. As described by Bard and Faulkner, SWV contrastingly measures current two times per cycle in the forms of a forward and reverse current [2]. These values are subtracted from one another to yield a difference current, often resulting in better analyte sensitivity due to suppressed background.

Varying concentrations of TNT are first titrated in the buffer system using the film assemblies as a working electrode and subsequently measured electrochemically with SWV (Fig. 10.5). Although bare Au displays a high current output at 100 μM TNT concentrations, only one peak is resolved due to the high background signal. Dissolved molecular oxygen (O_2) in the buffer attributes significantly to background signal at high TNT concentrations, due to the overlapping reduction potentials of dissolved oxygen and TNT (~0.4 V vs. Ag/AgCl). Below 10 μM TNT at a bare Au surface, no reduction peaks are observed because the TNT signal is masked by high background species.

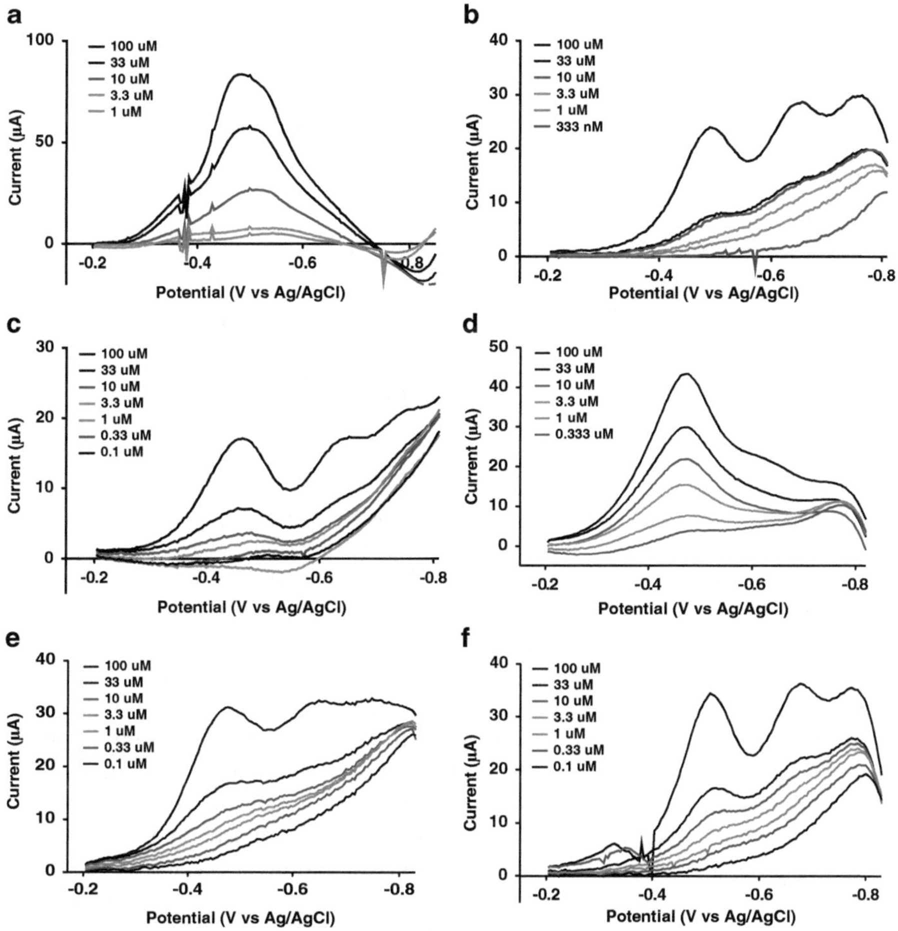

Fig. 10.5 Titrations of TNT measured via square wave voltammetry at (**a**) bare Au, (**b**) MUA-PLL, (**c**) 5 nm AuNP film assembly, (**d**) 15 nm AuNP film assembly, (**e**) 30 nm AuNP film assembly, and (**f**) 50 nm AuNP film assembly

For the MUA-PLL organic film, background current is almost entirely suppressed, generating excellent three-peak resolution at high TNT concentrations, each peak representative of the $6e^-/6H^+$ reduction from a nitro to amine group (Fig. 10.5b). However, at the lower concentrations, current output decreases significantly, most likely due to the weak electron tunneling effects through organic insulating molecules. It is also interesting to note that the peak height intensity is smaller with every subsequent reduction peak, as the stability of each reduced species becomes increasingly weaker. Colloidal AuNP films exhibit all three reduction peaks at high TNT concentrations (Fig. 10.5c–f). The first reduction peak is especially evident below 1 μM TNT in the 5 and 15 nm AuNP assemblies, while the 30 and 50 nm AuNP films exhibit visible peak heights down to 1 μM.

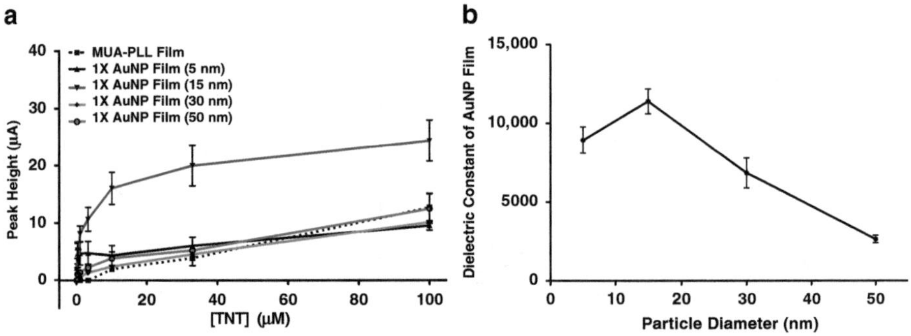

Fig. 10.6 AuNP film assembly performance. (**a**) TNT titrations as a function of the first reduction peak height and (**b**) calculation of a theoretical dielectric constant for each AuNP film

Because sensitivity increases at the AuNP films, the electron tunneling mechanism through the films is refuted while the electron hopping mechanism among particles is supported. The electrons are being shuttled distances on the order of 100 nm at the AuNP assemblies in comparison to less than a 5 nm distance with bulk Au and organic SAM films. It is also observed that the 15 nm AuNP films exhibit the greatest sensitivity (Fig. 10.6). It was originally hypothesized that the 5 nm AuNP films would display the highest sensitivity because the particles possess the largest relative surface area and a higher relative concentration of high-energy surface electrons. The high surface energy of the 5 nm AuNP film assemblies would therefore shuttle electrons more efficiently. However, the 15 nm AuNP films exhibit the highest saturated signal and sensitivity, offering a detection limit of 336 nM. The 5 nm AuNP films also demonstrate high nanomolar limits of detection, but with lower sensitivity and signal-to-noise. Further investigations are necessary to identify which factors (stacking of particles in film, assembly, film defects, etc.) cause the improved sensitivity in the 15 nm AuNP films.

To gain a better understanding of why the 15 nm AuNP films exhibit the highest sensitivity, the capacitance behavior of each film is examined. The 5 and 15 nm AuNP films exhibit high double layer capacitance values, indicating they store charge effectively. This value can be calculated by modeling these films after an electrical double-plate capacitor, as:

$$C_{dl} = \frac{I_{Tot}}{2 * \upsilon} = \frac{A * \in * \in_0}{d},$$

where C_{dl} is the double-layer capacitance in Farads, I_{Tot} is the total current or current thickness value (A), υ is the cyclic voltammetry scan rate (V/s), A is the electrode area (cm^2), d is the film thickness (m), ε_0 is the vacuum permittivity of free space constant, and ε is the dielectric constant of the material [2, 18, 26, 27]. The values I_{Tot} and d, are measured experimentally via cyclic voltammetry and profilometry, respectively. After substituting these values into the above equation, the value for dielectric constant can be solved. It is interesting to note that the 15 nm AuNP films exhibit the highest dielectric constant, which suggests better charge storing capabilities than any other particle assembly (Fig. 10.6b). The 5 and 30 nm AuNP films exhibit similar dielectric constant values, while the 50 nm AuNP film exhibits the lowest dielectric constant. This result suggests that there may be a size-dependent aspect for particle packing efficiency at this type of substrate.

10.3.4 Catalytic Properties of AuNP Film Assemblies

In conventional electrochemistry, higher free energy corresponds to more negative potentials, while a lower free energy corresponds to more positive potentials. If a reduction occurs at more positive potentials, a smaller quantity of energy is required to electrochemically reduce the analyte of interest. There are previously reported studies elucidating size-dependent catalytic applications of AuNPs, but TNT electrocatalysis at AuNP films has never been investigated [28–31]. Also interesting to note is that there are conflicting reports of AuNP size-dependence, suggesting that catalysis may be application-specific. It was hypothesized that the 5 nm AuNP film would generate the largest electrocatalytic activity (i.e. largest positive shift in reduction potential) because these particles contain the highest amount of relative surface atoms compared to the larger particles. The onset potential for TNT reduction is utilized in order to probe for the electrocatalytic activity of these films, because at this potential, the analyte begins its electrochemical reduction process. By fitting a line to the peak shape of the first TNT reduction peak and subsequently solving for the value of the x-intercept, onset reduction potential of a constant concentration of TNT can be calculated at each AuNP film assembly (Fig. 10.7). At 5, 15, and 30 nm AuNP film assemblies, the onset reduction potentials of TNT are similar, shifting approximately 70 mV more positive relative to bare Au and MUA-PLL negative control electrodes. The 50 nm AuNP film shows an insignificant shift in onset potential and thus, zero electrocatalytic activity. Between 5 and 30 nm, colloidal AuNP assemblies act as catalysts for electrochemical TNT reduction because of their high surface energy due to their monodisperse nature after assembly. Unlike the predicted behavior, the 5, 15, and 30 nm AuNP assemblies all exhibit similar shifts in onset reduction potential, suggesting a threshold for electrocatalytic activity may be achieved by nanoparticle films of less than 30 nm assembled on gold substrates.

10.4 Conclusions

This chapter describes the facile platform sensor development for TNT detection using cheap, stable AuNPs films produced by simple electrostatic adsorption. The particles demonstrate two functions: creating a blocking layer similar to a

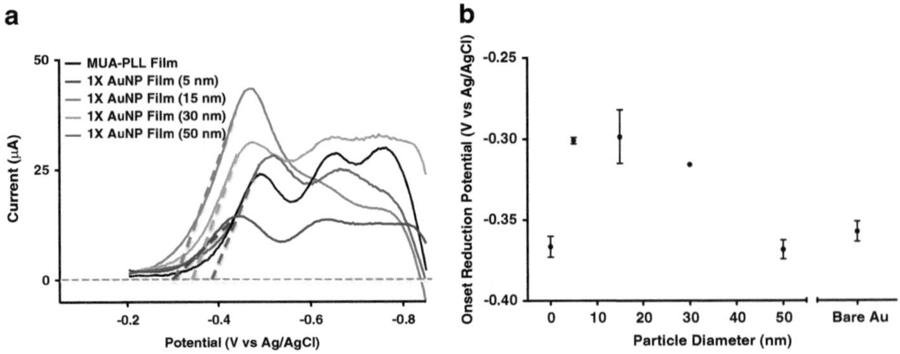

Fig. 10.7 Catalytic performance of AuNP film assemblies in the presence of 100 μM TNT. (**a**) Square wave voltammograms at various assemblies and visual representation of calculation of onset potential for the first reduction peak of TNT. A line is fitted to the peak shape of reduction and the value of the x-intercept denotes the onset potential. (**b**) Plot of onset reduction potential of first TNT peak as a function of particle size (0 diameter particle size corresponds to MUA-PLL organic film)

SAM to shield oxygen redox signal effects but to also maintain favorable electron transfer like the bulk Au counterpart. At 5, 15, and 30 nm AuNP film assemblies, electrocatalysis is evident by the 70 mV shifts in onset reduction potential of the first TNT peak. An optimized performance also emerges concerning nanoparticle diameter, as the 15 nm AuNP films exhibit the highest TNT sensitivity, lowest limits of detection, most positive onset reduction potential, and largest calculated dielectric constant. Although further investigation is needed, it is hypothesized that the stable 15 nm particles can pack more efficiently at the gold surface compared to AuNPs of alternative sizes. Additionally, there may be a threshold size related to nanoparticle stability at these films, as the 50 nm AuNPs show zero catalytic activity. Nonetheless, the innovation inspired by utilizing cheap materials has the ability to streamline sensor development for a wide variety of analytes, in addition to offering novel catalysts at solid interfaces and suspended in synthetic reactions.

Acknowledgements The authors wish to thank NATO for providing the opportunity and funding to present this work in Antalya, Turkey and Alexis Wong for capturing TEM images on the Tecnai Osiris (NSF EPS 1004083) at Vanderbilt University using VINSE facilities.

References

1. Moore DS (2004) Instrumentation for trace detection of high explosives. Rev Sci Instrum 75:2499–2512
2. Bard AJ, Faulkner LR (2001) Electrochemical methods: fundamentals and applications, 2nd edn. Wiley, New York

3. Nath S, Ghosh SK, Kundu S, Praharaj S, Panigrahi S, Pal T (2006) Is gold really softer than silver? HSAB principle revisited. J Nanopart Res 8:111–116
4. Bain CD, Troughton EB, Tao YT, Evall J, Whitesides GM, Nuzzo RG (1988) Formation of monolayer films by the spontaneous assembly of organic thiols from solution onto gold. J Am Chem Soc 111:321–335
5. Love JC, Estroff LA, Kriebel JK, Nuzzo RG, Whitesides GM (2005) Self-assembled monolayers of thiolates on metals as a form of nanotechnology. Chem Rev 105:1105–1169
6. Nuzzo RG, Allara DL (1983) Adsorption of bifunctional organic disulfides on gold surfaces. J Am Chem Soc 105:4481–4483
7. Li G, Miao P (2013) Electrochemical analysis of proteins and cells. Springer, Berlin
8. Kwon SJ, Yang H, Jo K, Kwak J (2008) An electrochemical immunosensor using p-aminophenol redox cycling by NADH on a self-assembled monolayer and ferrocene-modified Au electrodes. Analyst 133:1599–1604
9. Avila A, Gregory BW, Niki K, Cotton TM (2000) An electrochemical approach to investigate gated electron transfer using a physiological model system: cytochrome c immobilized on carboxylic acid-terminated alkanethiol self-assembled monolayers on gold electrodes. J Phys Chem B 104:2759–2766
10. Chi Q, Zhang J, Andersen JET, Ulstrup J (2001) Ordered assembly and controlled electron transfer of the blue copper protein azurin at gold (111) single-crystal substrates. J Phys Chem B 105:4669–4679
11. Davis KL, Drews BJ, Yue H, Waldeck DH, Knorr K, Clark RA (2008) Electron-transfer kinetics of covalently attached cytochrome c/SAM/Au electrode assemblies. J Phys Chem C 112:6571–6576
12. Flemming BD, Proporski S, Bond AM, Martin LL (2008) Electrochemical quartz crystal microbalance study of azurin adsorption onto an alkanethiol self-assembled monolayer on gold. Langmuir 24:323–327
13. Fujita K, Nakamura N, Ohno H, Leigh BS, Niki K, Gray HB, Richards JH (2004) Mimicking protein-protein electron transfer: voltammetry of *Pseudomonas aeruginosa* azurin and the *Thimus thermophilus* Cu-A domain at omega-derivatized self assembled-monolayer gold electrodes. J Am Chem Soc 126:13954–13961
14. Nakano K, Yoshitake T, Yamashita Y, Bowden EF (2007) Cytochrome c self-assembly on alkanethiol monolayer electrodes as characterized by AFM, IR, QCM, and direct electrochemistry. Langmuir 23:6270–6275
15. Niki K, Hardy WR, Hill MG, Li H, Sprinkle JR, Margoliash E, Fujita K, Tanimura R, Nakamura N, Ohno H, Richards JH, Gray HB (2003) Coupling to Lysine-13 promotes electron tunneling through carboxylate-terminated alkanethiol self-assembled monolayers to cytochrome c. J Phys Chem B 107:9947–9949
16. Song S, Clark RA, Bowden EF, Tarlov MJ (1993) Characterization of cytochrome c/alkanethiolate structures prepared by self-assembly on gold. J Phys Chem 97:6564–6572
17. Yokoyama K, Leigh BS, Sheng Y, Niki K, Nakamura N, Ohno H, Winkler JR, Gray HB, Richards JH (2008) Electron tunneling through *Pseudomonas aeruginosa* azurins on SAM gold electrodes. Inorg Chim Acta 361:1095–1099
18. Loftus AF, Reighard KP, Kapourales SA, Leopold MC (2008) Monolayer-protected nanoparticle film assemblies for controlling interfacial and adsorption properties in protein monolayer electrochemistry. J Am Chem Soc 130:1649–1661
19. Vargo MV, Gulka CP, Gerig JK, Manieri CM, Dattelbaum JD, Marks CB, Lawrence NT, Trawick ML, Leopold MC (2010) Distance dependence of electron transfer kinetics for azurin protein adsorbed to monolayer protected nanoparticle film assemblies. Langmuir 26:560–569
20. Schmidt AR, Nguyen NDT, Leopold MC (2013) Nanoparticle film assemblies as platforms for electrochemical biosensing-factors affecting the amperometric signal enhancement of hydrogen peroxide. Langmuir 29:4574–4583
21. Chua CK, Pumera M, Rulíšek L (2012) Reduction pathways of 2,4,6-trinitrotoluene: an electrochemical and theoretical study. J Phys Chem C 116:4243–4251

22. Goh MS, Pumera M (2011) Graphene-based electrochemical sensor for detection of 2,4,6-trinitrotoluene (TNT) in seawater: the comparison of single-, few-, and multilayer graphene nanoribbons and graphite microparticles. Anal Bioanal Chem 399:127–131
23. Ma H, Yao L, Li P, Ablikim O, Cheng Y, Zhang M (2014) Highly sensitive and selective fluorometric/electrochemical dual-channel sensors for TNT and DNT explosives. Chem Eur J 20:11655–11658
24. de Sanoit J, Vanhove E, Mailley P, Bergonzoa P (2009) Electrochemical diamond sensors for TNT detection in water. Electrochim Acta 54:5688–5693
25. Yu Y, Cao Q, Zhou M, Cui H (2013) A novel homogeneous label-free aptasensor for 2,4,6-trinitrotoluene detection based on an assembly strategy of electrochemiluminescent graphene oxide with gold nanoparticles and aptamer. Biosens Bioelectron 43:137–142
26. Finklea HO (1996) Electroanalytical chemistry. Marcel Dekker, New York
27. Wang H, Pilon L (2011) Accurate simulations of electric double layer capacitance of ultramicroelectrodes. J Phys Chem C 115:16711–16719
28. Lin C, Tao K, Hua D, Ma Z, Zhou S (2013) Size effect of gold nanoparticles in catalytic reduction of p-nitrophenol with $NaBH_4$. Molecules 18:12609–12620
29. Zhou X, Xu W, Liu G, Panda D, Chen P (2010) Size-dependent catalytic activity and dynamics of gold nanoparticles at the single-molecule level. J Am Chem Soc 132:138–146
30. Choi Y, Choi MJ, Cha SY, Kim YS, Cho S, Park Y (2014) Catechin-capped gold nanoparticles: green synthesis, characterization, and catalytic activity toward 4-nitrophenol reduction. Nanoscale Res Lett 9:103–110
31. Brülle T, Ju W, Niedermayr P, Denisenko A, Paschos O, Schneider O, Stimming U (2011) Size-dependent electrocatalytic activity of gold nanoparticles on HOPG and highly boron-doped diamond surfaces. Molecules 16:10059–10077

Chapter 11
Nanoindentation Investigations of $(As_2Se_3)_{1-x}$: Sn_x and $(As_4S_3Se_3)_{1-x}$: Sn_x Glasses

D.V. Harea, E.E. Harea, O.V. Iaseniuc, and M.S. Iovu

Abstract Experimental results on some physical and optical properties of $(As_2Se_3)_{1-x}$:Sn_x and $(As_4S_3Se_3)_{1-x}$:Sn_x ($x = 0$–10 at.%) glasses and amorphous films ($d \sim 2.0$ μm) are presented. The bulk chalcogenide glasses are studied by *X*-ray diffraction spectroscopy and nanoindentation methods. It is established that the addition of these amounts of tin ($x = 0$–10 at.%) does not lead to significant changes in the physical properties of the glass, such as values of stress and Young's modulus related to the modification of the density and compactness. The XRD measurements show that the Sn impurities in the $(As_4S_3Se_3)_{1-x}$:Sn_x do not significantly change the shape of the first sharp diffraction peak (FSDP) of the X-ray diffraction patterns either; the intensity and the position of the FSDP nonmonotonically depend on the Sn concentration. It has been found that the addition of these amounts of tin in $(As_4S_3Se_3)_{1-x}$:Sn_x does not lead to significant changes in the glass physical properties, such as values of stress and Young's modulus related to the modification of the density and compactness. The study of the photoplastic effect is performed in situ, with illumination of the bulk and thin film samples during indentation as well as their indentation after illumination with a green laser ($\lambda = 532$ nm) at a power of $P = 50$ mV/cm^2. The hardness is calculated from load-displacement curves by the Oliver-Pharr method. A sharp increase in hardness is registered if the tin concentration exceeds a value of 3–4 % Sn. The hardness H of $(As_2Se_3)_{1-x}$:Sn_x films varies between 115 and 130 kg/mm^2. It is found that the hardness H of amorphous thin films is generally higher than the hardness of bulk samples with the same chemical composition. In this study, we are focused on the mechanical characteristics of high-purity As_2Se_3:Sn_x thin films.

D.V. Harea (✉) • E.E. Harea • O.V. Iaseniuc • M.S. Iovu
Institute of Applied Physics, Academy of Sciences of Moldova, Academiei str. 5, Chisinau MD-2028, Republic of Moldova
e-mail: dyanaharea@gmail.com

T.A. Camesano (ed.), *Nanotechnology to Aid Chemical and Biological Defense*, NATO Science for Peace and Security Series A: Chemistry and Biology,
DOI 10.1007/978-94-017-7218-1_11

11.1 Introduction

Chalcogenide glasses (CGs), such as As_2S_3, As_2Se_3, As_2Te_3, As-S-Se, As-Sb-S(Se), As-Ge-S(Se), represent a new class of advanced materials exhibiting attractive combinations of properties, such as high strength/hardness and excellent wear/corrosion resistance [1–4], excellent optical properties (high refractive index, high transmittance in the near IR and IR regions of the spectrum). This makes them good candidates for photonic and optoelectronic applications (such as optical elements and memories, optical sensors, nonlinear optical devices, holographic elements, IR telecommunications, biosensing, signal processing, and photonic applications [5–7]). The effect of light-induced photostructural transformations is characteristic of many amorphous chalcogenides films, and they have served as a base of many applications in photonics and optoelectronics, especially as inorganic photoresists for submicron technology [8, 9].

These distinguished properties are primarily due to a disordered atomic arrangement in the amorphous materials, resulting in the absence of grain boundaries and defects in the microstructure. While the nonequilibrium nature of amorphous materials offers outstanding properties, it also presents significant challenges in the processing of these materials.

It is even more interesting to study the chalcogenides doped with metal impurities, which alter the physicochemical, electrical, and optical properties. It was shown that the Sn impurity introduced in the As_2Se_3, AsSe, and Sb_2S_3 glass network reduces the photodarkening effect [10–13]. Tin-containing chalcogenides showed the some deviation in behavior due to modification in the local ordering of the host chalcogenide matrix after the tin incorporation, which was observed in the bulk glasses as well as in the corresponding thin films.

For numerous applications, the mechanical behavior is one of important characteristics of the material [14–16]. Moreover, the photoplastical effect in some amorphous films (As-Se, Ss-S-Se) that occurs under band-gap illumination is described. This effect is considered electronic rather than thermal and consists of two parts: negative (decreasing in viscosity) and positive (increasing in viscosity); it occurs only under illumination. In the opinion of the authors, the first part of the photoplastic effect is general for CGs and causes all photostructural phenomena in amorphous materials.

The photoplastical effect is evident from the hardening of illuminated materials [17] as well as their softening [18]. The photoplastical effect depends on a number of factors, such as radiation power, temperature, and wavelength. Studies of the spectral dependence of photoplasticity showed a maximum effect under illumination of samples with a wavelength close to the band gap value [19]. Photoplasticity in chalcogenide films was studied in [20–22], where the effect was observed during illumination of samples with a wavelength comparable to the band gap value. The nature of the photoplastical effect remains poorly understood. The effect was attributed to thermal expansion of the film due to the absorption of exciting light as well as recombination of electrons and holes under photoexcitation. The properties of glass can be varied and regulated over an extensive range by modifying the composition and production techniques. It was shown that the addition of a tin impurity in amorphous As_2Se_3 films can provide a pronounced effect on electrical, transport properties, optical and photoinduced phenomena [23–27].

In this study, we have combined the experimental investigations of X-ray diffraction (XRD) measurements and the mechanical properties of bulk and thin films of $(As_2Se_3)_{1-x}$:Sn_x and $(As_4S_3Se_3)_{1-x}$:Sn_x ($x = 0 \div 10$ at.% Sn) chalcogenides.

11.2 Experimental

The bulk CGs $(As_2Se_3)_{1-x}$:Sn_x and $(As_4S_3Se_3)_{1-x}$:Sn_x ($x = 0 \div 10$ at.% Sn) glasses were prepared from the elements of 6 N (As, S, Se, Sn) purity by a conventional melt quenching method. The starting components $As_4S_3Se_3$ and Sn were mixed in quartz ampoules and then evacuated to a pressure of $P \sim 10^{-5}$ Torr, sealed and heated to a temperature of $T = 900$ °C at a rate of 1 °C/min. The quartz tubes were held at this temperature for 48 h for homogenization and then slowly quenched in a furnace. For nanoindentation and optical measurements of the bulk samples, planar parallel plates with a thickness of about $d = 2$–3.5 mm were prepared and polished. Experiments were carried out at room temperature. Indentation was performed with CSM Indentation Testers (Ultra Nano, Nano and Micro; with max load $P = 100$ mN; load resolution of 0.001 μN) with software. Hardness (Hv) and Young's modulus (E) were automatically computed. The X-ray diffraction (XRD) measurements were performed on a DRON-UM1 diffractometer using *Fe-Kα* radiation ($\lambda = 1.93604$ Å) with a Mn filter by the $\theta/2\theta$ scanning method. Thin film samples with a thickness of $d = 0.5$–3 μm were prepared by vacuum flash thermal evaporation of the synthesized initial CG onto glass substrates held at $T_{subs} = 100$ °C.

11.3 Results and Discussion

11.3.1 X-Ray Diffraction Measurements

Crystalline semiconductors are characterized by the *ordering on distance*, e.g., by the existence of a correlation between two atoms situated at any distance in the network. The noncrystalline semiconductors preserve a *short range order (SRO)* in a range of 0.3–0.5 nm and are characterized by the existence of a correlation in the first atomic coordination sphere. In the case of dominantly covalently bounded amorphous solids, SRO is described by local coordination polygons, e.g., pyramidal AsS_3 in As_2S_3. In many amorphous semiconductors, such as CGs, this order is shown at longer distances, the so-called *medium range order (MRO)* which extends up to a range of 0.5–1.0 nm. For disordered semiconductors, a *long range order (LRO)* is absent, and this means that the amorphous chalcogenides have no translational symmetry.

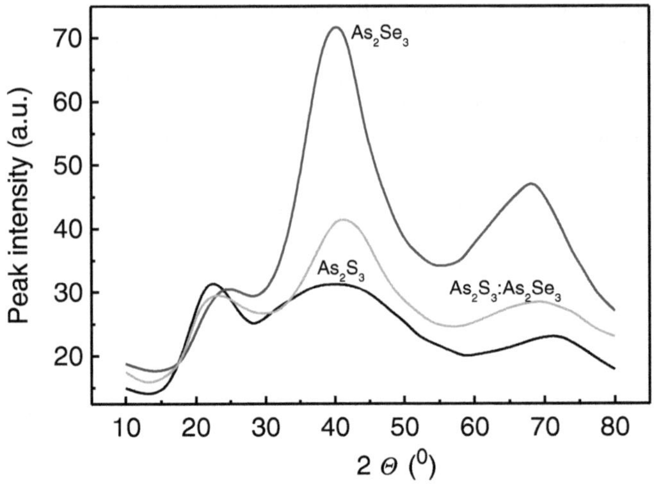

Fig. 11.1 X-ray diffraction patterns for As_2S_3, s_2Se_3, and $As_4S_3Se_3$ glasses

The X-ray diffraction technique is a nondestructive method which gives information from the atomic scale range; it is a widely used method for investigations of disordered semiconductors with multilayered structures [28]. Using the X-Ray diffraction method, the diffraction patterns were obtained in the range of 2θ diffraction angles of 10–80° (θ is the Bragg angle) for CGs As_2S_3, As_2Se_3, $As_4S_3Se_3$, and $As_4S_3Se_3$:Sn_x (x = 0.01, 0.02, 0.04, 0.06, 0.07, and 0.10).

Figure 11.1 shows the angular distribution of X-Ray diffraction intensity for As_2S_3, As_2Se_3, and $As_4S_3Se_3$ chalcogenides glasses. The position of the first sharp diffraction peak (FSDP) for As_2S_3 is $2\theta = 22.47°$ and increases to $2\theta = 24.60°$ for As_2Se_3. For the intermediate composition of $As_4S_3Se_3$, the maximum of the FSDP is located at $2\theta = 23.00°$. These spectra represent a sum of diffraction patterns of is structural vitreous As_2S_3 and As_2Se_3 with three broad lines of diffractograms, which are similar to the unvelope of the rounded lines of the spectra of crystalline As_2S_3 and As_2Se_3. It can be assumed that, in the microcrystalline state of the investigated glasses, there are domains with an ordered structure with dimensions of about 15–20 Å. Previously, an analogy between the structure of vitreous and crystalline states of As_2S_3 was vindicated by short-range order investigations – interatomic distances and coordination numbers – with the addition of the radial distribution function [28].

It was established that Van der Waals forces with a reduced covalent component act between As_2S_3 and As_2Se_3 layers. The interaction forces between the layers are hundred times weaker than the binding forces between the layers. According to [28], the structure of the glasses represents an interlinking of As-S_3 and As-Se_3 pyramids that form rings with six units. The arsenic atoms are situated at the top of the pyramid, while the chalcogen atoms form the basis. As was shown, the crystalline semiconductors are characterized by a LRO; for example, there is a good correlation between the position of the each two atoms in the network, which is not found in noncrystalline semiconductors [29, 30].

Noncrystalline semiconductors exhibit only a SRO, which involves individual atoms in the first coordination sphere. Since the range order in CGs can be extended to several interatomic distances, a new concept of the average order was introduced (MRO).

Comprehensive studies of the FSDP of vitreous As_2S_3 and As_2Se_3 show that they have a similar structure [28]. According to [28], the first coordination spheres (first order neighbor position) of the central atom in the structure is $r_1 = 2.414$ Å for As_2Se_3, and $r_1 = 2.306$ Å for As_2S_3, respectively. The second coordination spheres (second order neighbor position) of the central atom in the structure is $r_1 = 3.625$ Å for As_2Se_3, and $r_1 = 3.475$ Å for As_2S_3, respectively.

The Sn concentration in the mixed glasses $As_4S_3Se_3$:Sn_x does not significantly change the shape of the FSDP of the X-ray diffraction patterns (Fig. 11.2). In general, the diffraction patterns of the $As_4S_3Se_3$:Sn_x glasses are similar and form three wide lines, the maxima of which correspond to the interlayer distances *d*–4.8–2.8–1.7 Å. As in the case of As_2Se_3:Sn_x [30], the angular position of the FSDP in $As_4S_3Se_3$:Sn_x slightly depends on the Sn concentration. Figures 11.3 and 11.4 show the angular position and the intensity of the FSDP in the X-ray diffraction patterns of As_2S_3 (1), As_2Se_3 (2), and $As_4S_3Se_3$ (3), and the FSDP in the X-ray diffraction patterns of the $(As_4S_3Se_3)_{1-x}$:Sn_x glasses.

Some parameters of the X-ray diffraction patterns of the investigated CGs are listed in Table 11.1. Figures 11.5 and 11.6 show the dependences of the peaks intensities vs. concentrations of tin of the three diffraction peaks of CG $As_4S_3Se_3$:Sn_x situated at 2θ–22.86–23.67°, 2θ–40.34–42°, and 2θ–62–70°, and the

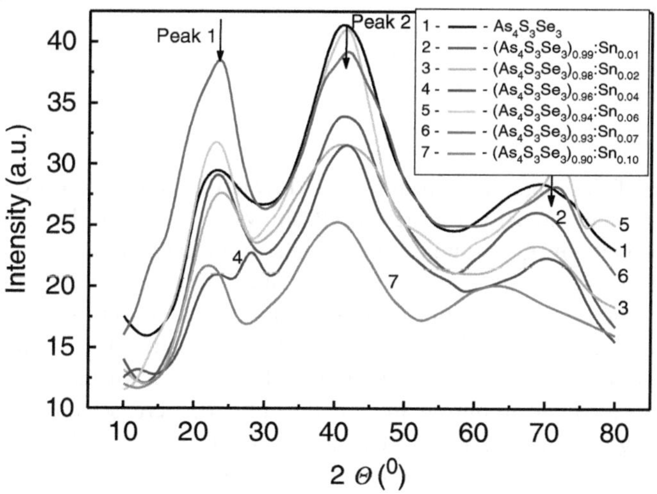

Fig. 11.2 X-ray diffraction patterns for $(As_4S_3Se_3)_{1-x}$:Sn_x glasses; *x*, at % Sn: (1) 0, (2) 1.0, (3) 2.0, (4) 4.0, (5) 6.0, (6) 7.0, and (7) 10.0

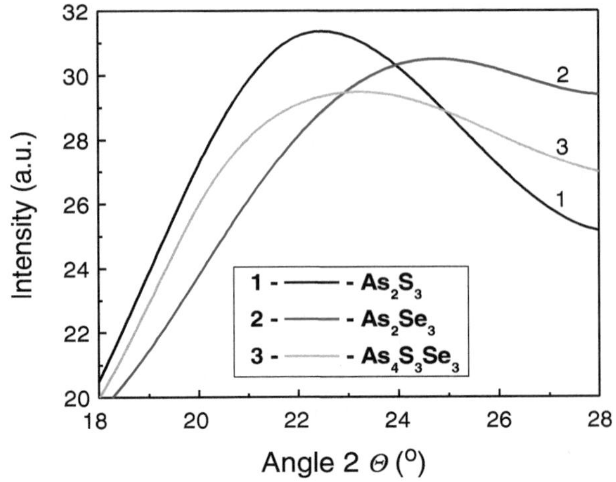

Fig. 11.3 FSDP in the X-ray diffraction patterns of As_2S_3 (1), As_2Se_3 (2), and $As_4S_3Se_3$ (3)

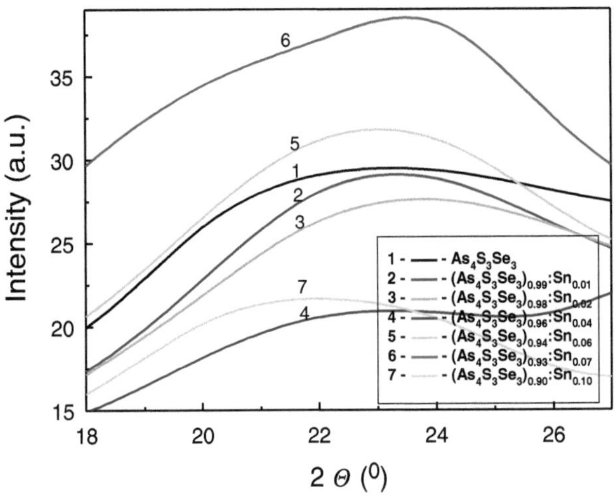

Fig. 11.4 FSDP in the X-ray diffraction patterns of $(As_4S_3Se_3)_{1\text{-}x}$:$Sn_x$; *x*, at % Sn: (1) 0, (2) 1.0, (3) 2.0, (4) 4.0, (5) 6.0, (6) 7.0, and (7) 10.0

dependence of the angular position of the diffraction peaks vs. tin concentration, respectively. The intensity of the FDSP shows a nonlinear behavior with the different amounts of doping with Sn. The maximum intensity is achieved at about 6 at.% Sn in $As_4S_3Se_3$:Sn_x, while in the case of As_2Se_3:Sn_x the maximum intensity is situated at the tin concentration of about 2 at.% Sn. A similar behavior was found for the second and third diffraction peaks. According to [30], when Sn is added in As_2Se_3 or As_2S_3 ChGs, due to the tetrahedral disposal of the sp3 bonds in the chalcogens, the dopant atom implanted in the network increases the thickness of the layered configuration as revealed by a significant shift of the FSDP towards lower angles. This implantation corresponds to the introduction of the structural units of the $SnSe_2$ or SnS_2 type in the glass network; the same fact was confirmed by the Mössbauer spectroscopy experiments [11].

Table 11.1 Parameters of the X-ray diffraction patterns of the studied As_2S_3, As_2Se_3, and $(As_4S_3Se_3)_{1-x}$:Sn_x CGs. *For the* $(As_4S_3Se_3)_{0.96}$:$Sn_{0.04}$*glass composition, an additional peak at* $2\theta = 28.20$ *was observed*

N/o	Glass composition	Position of peak 1, 2θ	Intensity of peak 1, a.u.	Position of peak 2, 2θ	Intensity of peak 2, a.u.	Position of peak 3, 2θ	Intensity of peak 3, a.u.
1	As_2S_3	22.32	31.36	40.50	31.07	71.78	22.92
2	As_2Se_3	24.35	30.52	40.50	71.54	68.25	46.81
3	$As_4S_3Se_3$	22.86	29.43	41.01	41.14	69.28	28.37
4	$(As_4S_3Se_3)_{0.99}$:$Sn_{0.01}$	23.39	29.06	41.53	33.92	69.15	26.09
5	$(As_4S_3Se_3)_{0.98}$:$Sn_{0.02}$	23.67	27.54	41.14	31.64	68.61	23.51
6	$(As_4S_3Se_3)_{0.96}$:$Sn_{0.04}$	22.86	20.93	41.53	31.49	70.62	22.38
7	$(As_4S_3Se_3)_{0.94}$:$Sn_{0.06}$	23.00	31.64	41.53	41.14	72.08	30.35
8	$(As_4S_3Se_3)_{0.93}$:$Sn_{0.07}$	23.39	38.33	41.81	39.17	71.55	28.15
9	$(As_4S_3Se_3)_{0.90}$:$Sn_{0.10}$	21.93	21.69	40.34	25.26	62.75	20.17

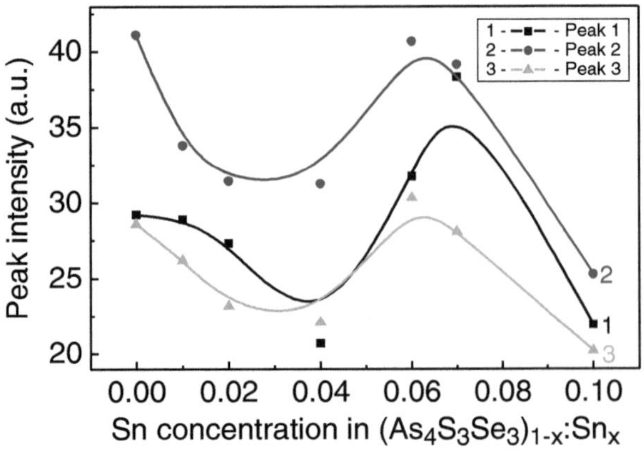

Fig. 11.5 Dependence of the angular position of the diffraction peaks vs. concentration of tin in some $As_4S_3Se_3:Sn_x$ CGs

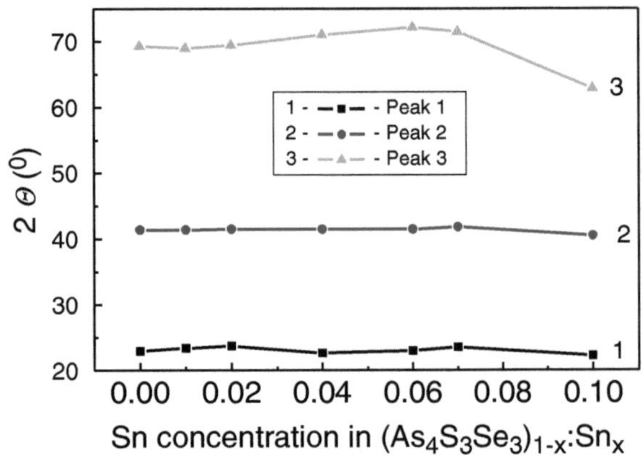

Fig. 11.6 Dependence of the angular position of the diffraction peaks vs. concentration of tin in the some $As_4S_3Se_3:Sn_x$ CGs

11.3.2 Physical Properties and Photoplastic Effect in $As_2Se_3:Sn_x$ Glasses

The physical properties (density, hardness, glass transition temperature, electrical conductivity, etc.) of amorphous As-Se films have been previously reviewed by Borisova [31]. The investigation of bulk, thin films, and illuminated thin film samples was performed using a NHT CSM nanohardness tester. The hardness was calculated from load-displacement curves by the Oliver-Pharr method. In-situ illumination of samples was carried with a green laser ($\lambda = 532$ nm) with a power of $P = 50$ mV/cm^2. To change the direction of the incident laser beam, an optical glass prism was used (Fig. 11.7).

The maximum indentation load was 5 mN; therefore, the maximal penetration depth did not exceed 15 % of the film thickness. The hardness was calculated using the expression,

$$H_B = (1570 \, * \, P)\,/\,l^2, \tag{11.1}$$

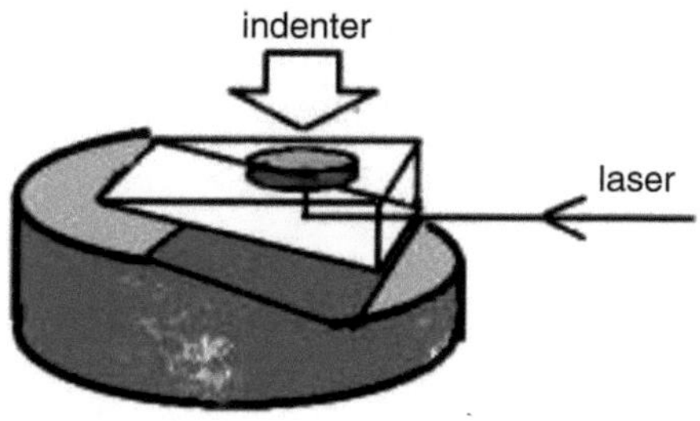

Fig. 11.7 Experimental set-up for investigation of the photoplastical effect in CGs

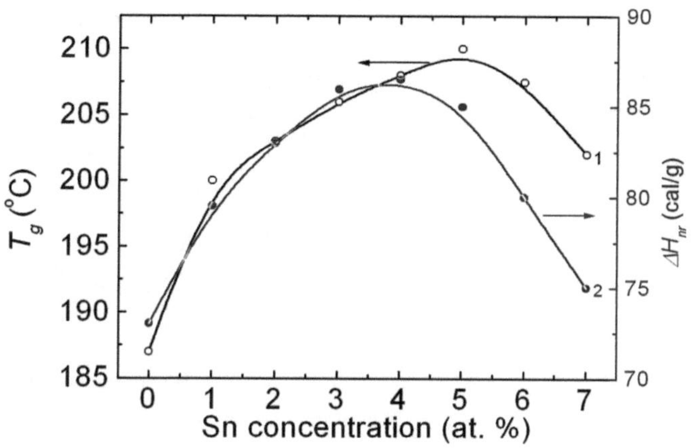

Fig. 11.8 Variations in $T_g(x)$ (1), $\Delta H_{nr}(x)$ for As_2Se_3:Sn_x glasses. The smooth lines are computer fitting

where P is the applied load and L is the height of the triangle of the remaining imprints [32]. For both the bulk samples and the amorphous thin films, the applied load was 10 g and the depth of deposited imprints did not exceed 20 % of the films thickness.

Figure 11.8 shows variations in glass transition temperature $T_g(x)$ for the bulk As_2Se_3:Sn_x glasses and the nonreversing heat $\Delta H_{nr}(x)$. At low concentrations of the Sn additive, the glass transition temperature T_g of the base glass increases with x, suggesting that the base glass becomes more bound. However, as x approaches 5 at.% Sn, the glass transition temperature T_g exhibits a threshold behavior [11].

The maximum in T_g suggests [11, 24] that the above threshold additives no longer form part of the base glass structure, e.g., they undergo nanoscale phase separation. These nanoscale phase separation effects have already been found in As_2Se_3 base glass ($x = 0$) and are apparently attributed to higher concentrations of Sn ($x > 5$ at.% Sn). Mössbauer spectroscopy results show the presence of a single line (A) feature in the spectra at low x (<3 at.%). The isomer-shift of this line has been previously assigned [33] to Sn that is tetrahedrally coordinated to 4 Se near-neighbors as in a $Sn(Se_{1/2})_4$ local structure [11]. Apparently, the introduction of a Sn additive in As_2Se_3 base glass contributes to the growth of $Sn(Se_{1/2})_4$ units and makes the base glass As-rich. The effect can occur in one of two modes: through the formation of either polymeric ethylene-like $As_{2}(Se_{1/2})_4$ units and/or monomeric

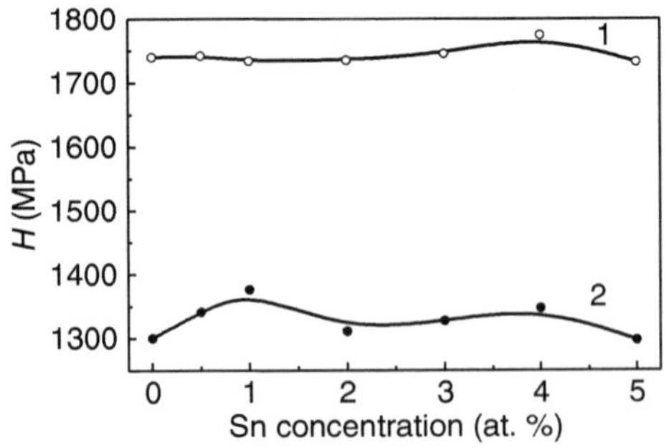

Fig. 11.9 Hardness *H* of $(As_2Se_3)_{1-x}$:Sn_x. samples, thin films (1) and bulk (2)

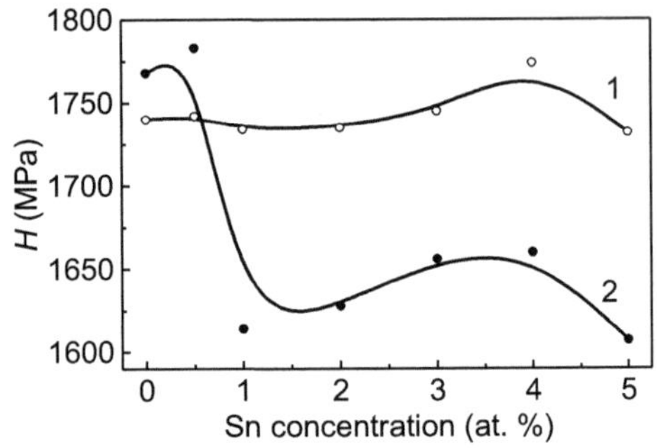

Fig. 11.10 Hardness *H* of $(As_2Se_3)_{1-x}$:Sn_x as-deposited (1) and illuminated (2) thin films

As_4Se_4 cages based on the Realgar structure. It is found that, in the base glass, the polymeric species are first nucleated at x slightly below 0.4, and monomeric ones are formed afterwards ($x > 0.42$). The presence of the latter fragments the backbone progressively at $x > 0.42$ leading to a reduction in T_g.

The experimental results of investigation of hardness for bulk and As_2Se_3:Sn_x amorphous thin films are presented in Figs. 11.9 and 11.10 and are in good agreement with experimental results obtained earlier by Borisova [31]. The hardness values of bulk samples depend on the Sn concentration in the As_2Se_3:Sn_x glasses and vary between 1300 and 1700 MPa.

The hardness *H* of the amorphous As_2Se_3:Sn_x thin films exhibits a nonmonotonic behavior in dependence on the Sn concentration and is higher than the hardness of bulk samples. An increase in hardness is registered at an impurity concentration of about 4 % Sn. The hardness *H* of $(As_2Se_3)_{1-x}$:Sn_x films varies between 1730 and 1780 MPa. For both bulk samples and thin films, the hardness is higher according to the data obtained in [31]. This fact can be attributed to some technological features in preparing the ChG. Under illumination, the hardness *H* of $(As_2Se_3)_{1-x}$:Sn_x decreases, and this decrease is specific especially of the doped samples (Fig. 11.10).

For investigation of microhardness and Young's modulus, a Nano-Indentation Tester (NIT) was used; it provides experimental data by indenting to depths at a nanometer–micron scale. This nanoindenter can be used to characterize organic, inorganic, soft or hard materials, and coatings. The mechanical characterization of the surface of bulk materials can also be performed, including metals, semiconductors, glasses, composites, and in vitro biomaterials.

The operating principle of the tester is as follows: an indenter diamond tip, normal to the sample surface, is driven into the sample by applying an increasing load ($P = 100$ mN) up to a certain preset value. The load is then gradually decreased until partial or complete relaxation of the material occurs. After the removal of the load, the diagonal length is measured (imprint d_1 and d_2). The offset of diagonal tip was <0.25 μm and the load resolution was 0.001 N. To avoid overlapping of surface stresses developed around neighboring indentations, the separation between indentation diagonals was kept more than ten times the diagonal length of indentation impressions.

The dimensions of both diagonals d made at a particular load P were measured, and the average diagonal d was calculated [14]. The value of microhardness H_V was computed from the $P(d)$ data using the standard well-known relation:

$$H_V = \frac{kP}{d_2} \tag{11.2}$$

where k is the geometrical conversion factor for the indenter used ($k = 0.1891$):

$$k = \frac{1}{g}\left(2\sin\frac{136^0}{2}\right) \tag{11.3}$$

where g is the acceleration of gravity, 136° is the angle between its opposite faces of pyramidal tip. The average values of indentation diagonal d and microhardness H_V for at least ten indentations were used in the analysis of indentation size effect and hardness measurements.

Figure 11.11 shows the Young's modulus and microhardness dependences for the bulk samples of As_2S_3, As_2Se_3, and for $(As_4S_3Se_3)_{1-x}$:Sn_x ($x = 0$–10 at.%) vs tin concentration. It was observed that the addition of tin in the host material initially leads to an increase in the microhardness up to 2.0 at.% Sn. A further increase in the Sn concentration decreases the microharness.

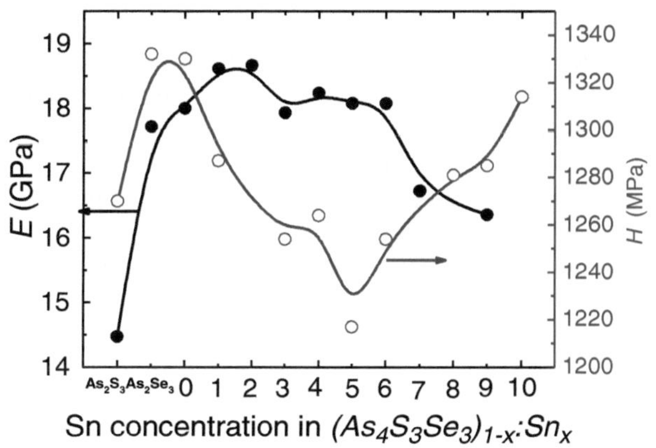

Fig. 11.11 Concentration dependence of Young's modulus E and microhardness H of the bulk samples As_2S_3, As_2Se_3, and $(As_4S_3Se_3)_{1-x}$:Sn_x

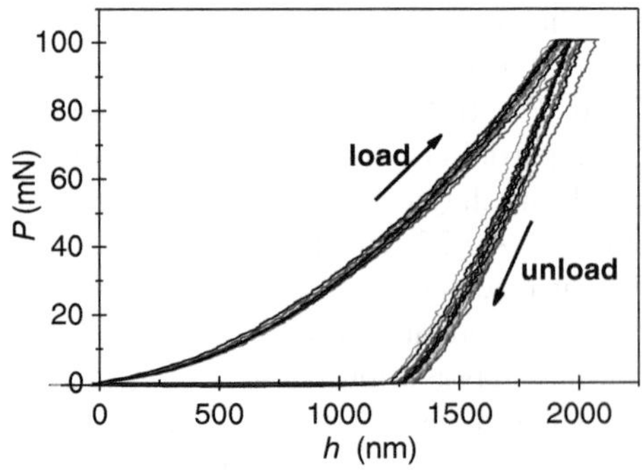

Fig. 11.12 Depth dependence of the load and unload values on chalcogenide samples of $(As_4S_3Se_3)_{1-x}:Sn_x$

In general, the hardness varies around one value $H_{middle} = 1282$ MPa. A similar behavior was observed for the system $(As_2S_3)_{1-x}:(Sb_2S_3)_x$, where a curve with the minimum at about 2–3 at.% of (Sb_2S_3) was obtained [34]. At the same time, the value of microhardness for vitreous As_2S_3 and As_2Se_3 better coincides with the values cited in [35], although the Young's modulus in our case is higher by an order. According to [34], the observed increase in the hardness with x in the $(As_2S_3)_{1-x}:(Sb_2S_3)_x$ system indicates structural changes, which lead to lower molecular or configuration mobility.

Figure 11.12 represents the depth dependence of the load and unload values for chalcogenide samples of $(As_4S_3Se_3)_{1-x}:Sn_x$. Similar dependences were obtained for amorphous As_2S_3 thin films prepared by thermal deposition in vacuum [36]. The author mentions this fact in these experiments because the hold segment is rather long and the unloading process is very fast, any further viscoplastic creep during unloading is negligible compared to the elastic recovery during the unloading procedure.

Figure 11.13 shows the surface morphology of the as-deposited amorphous $(As_2Se_3)_{0.90}:Sn_{0.10}$ thin film under the indentation load after (a) and before light illumination (b, upper patterns). For comparison, at the bottom of the figure, the pictures of the light-induced anisotropic plasticity in amorphous $As_{20}Se_{80}$ thin films are shown: (a) the image of the nonirradiated film surface and (b) the image of this surface by irradiated linearly polarized laser beam ($\lambda = 633$ nm) [37]. In both cases, after light irradiation, the plasticity of the investigated amorphous films increases. For explanation of this phenomenon, the authors of [36] proposed a mechanical model of anisotropic plasticity in CGs, according to which the anisotropic softening consists in the weakening of mechanical compliance in the orthogonal direction to the light polarization. Figure 11.14 represent the imprint on the surface of the diamond tip on a $(As_4S_3Se_3)_{1-x}:Sn_x$ sample at a load of $P = 100$ mN.

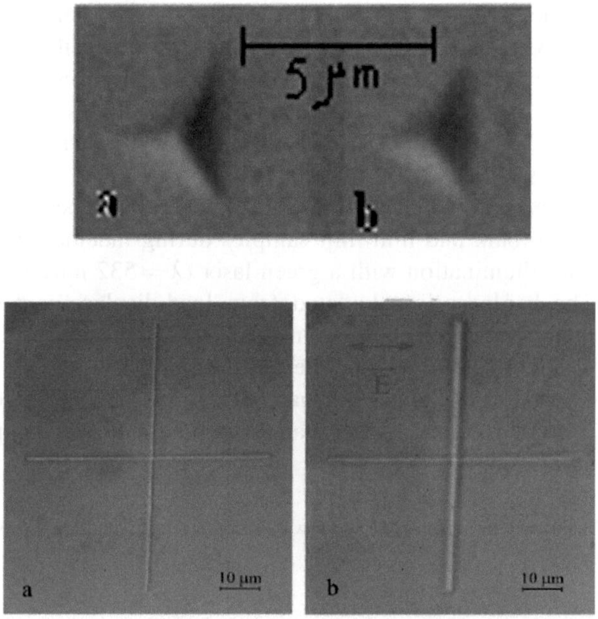

Fig. 11.13 Surface morphology of the as-deposited amorphous $(As_2Se_3)_{0.90}$:$Sn_{0.10}$ thin film under indentation load after (**a**) and before light illumination (**b**, upper patterns)

Fig. 11.14 Example of imprint on the surface of the diamond tip on bulk $(As_4S_3Se_3)_{1\text{-}x}$:$Sn_x$ samples at a load of $P = 100$ mN

11.4 Conclusions

The bulk of $(As_4S_3Se_3)_{1\text{-}x}$:$Sn_x$ ($x = 0$–10 at.% Sn) CGs have been investigated by X-ray diffraction and nanoindentation methods. It has been found that the addition of these amounts of tin does not lead to significant changes in the glass physical properties, such as values of stress and Young's modulus related to the modification of the density and compactness. The additions of Sn atoms in the CG matrix considerably decrease the elasticity of this material or increase the hardness at some tin concentrations. The XRD measurements have shown that the Sn impurities in the $(As_4S_3Se_3)_{1-x}:Sn_x$ do not significantly change the shape of the FSDP of the X-ray diffraction patterns either; the intensity and the position of the FSDP nonmonotonically depend on the Sn concentration. The addition of As_2Se_3 in As_2S_3 shifts the FSDP towards higher diffraction angles from $2\theta = 22.47^{\circ}$ for As_2S_3 up to $2\theta = 24.60^{\circ}$ for As_2Se_3. The FSDP intensity increases at about 5.0–7.0 at.% Sn in the $As_4S_3Se_3)_{1-x}:Sn_x$ glasses.

The study of the photoplastic effect has been performed in situ, with illumination of the bulk and thin film samples during indentation as well as their indentation after illumination with a green laser ($\lambda = 532$ nm) at a power of $P = 50$ mV/cm^2. The hardness is calculated from load-displacement curves by the Oliver-Pharr method. A sharp increase in hardness is registered if the tin concentration exceeds a value of 3–4 % Sn. The hardness H of $(As_2Se_3)_{1-x}:Sn_x$ films varies between 115 and 130 kg/mm^2. It has been found that the hardness H of amorphous thin films is generally higher than the hardness of bulk samples with the same chemical composition.

Acknowledgments The work was supported by the national project no. 11.817.05.03A.

References

1. Gerasimov AB, Chiradze GD, Kutivadze NG (2001) Semiconductors 35:72
2. Gerbreders A, Teteris J (2007) J Optoel Adv Mater 9(10):3164
3. Inoue A, Shen B, Takeuchi A (2008) Mater Trans 47:1275
4. Inoue A, Shen B, Nishiyama N (2008) In: Miller M, Liaw P (eds) Bulk metallic glasses. Springer, New York, pp 1–25
5. Inoue A (2009) Mater Sci Eng A 304–306:1
6. Andriesh AM, Iovu MS, Shutov SD (2002) J Optoel Adv Mater 4(3):631
7. Matsuda A, Mizuno H, Takayama T, Saito M, Ki-kuchi M (1974) Appl Phys Lett 25:411
8. Zakery A, Elliot SR (2007). Springer, Berlin
9. Sava F (2001) J Optoel Adv Mater 3(2):425
10. Koseki H, Odajima A (1982) Jpn J Appl Phys 21:424
11. Carlsson L, Svensson C (1970) J Appl Phys 41:1652
12. Tatarinova L (1983) The structure of amorphous solids and liquids. Nauka, Moscow, p 151
13. Lim YY, Munawar Chaudhri M (2006) Mech Mater 38:1213
14. Iovu M, Harea D, Colomeico E (2008) J Optoel Adv Mater 10(4):862
15. Iovu M, Boolchand P, Georgiev D (2005) J Optoel Adv Mater 7(2):763
16. Iovu M, Shutov S, Popescu M (2002). J NonCryst Solids 299:924
17. Trunov ML, Bilanich VS (2003) J Optoel Adv Mater 5(5):1085
18. Trunov ML (2008) J Phys D Appl Phys 41:074011, 9pp
19. Popescu M, Andries A, Ciumas V, Iovu M, Sutov S, Tiuleanu D (1996) Fizica sticlelor calcogenice. Editura Stiintifica Bucuresti – I.E.P, Stiinta
20. Popescu M, Tudorica F, Andriesh A, Iovu M, Shutov S, Bulgaru M, Colomeyko E, Malkov S, Verlan V, Leonovici M, Mihai V, Steflea M (1995) Buletinul Academiei de Stiinte a Republicii Moldova. Fizica si tehnica 3:3
21. M. Popescu (1996) In: Andriesh A, Bertolotti M (eds) Physics and applications of noncrystalline semiconductors in optoelectronics. pp 36, 215
22. Reinfelde M, Teteris J (2011) J Optoel Adv Mater 13(11–12):1531

23. Iovu MS, Harea DV, Cojocaru IA, Colomeico EP, Prisacari A, Ciorba VG (2007) J Optoel Adv Mater 9(10):3138
24. Iovu MS, Syrbu NN, Shutov SD, Vasiliev IA, Rebeja S, Colomeico E, Popescu M, Sava F (1999) Phys Status Solidi (a) 175(2):623
25. Iovu MS, Shutov SD, Toth L (1996) Phys Status Solidi B 195:149
26. Iovu MS, Shutov SD, Arkhipov VI, Adriaenssens GJ (2002) J Noncryst Solids 299&302:1008
27. Shchurova NN, Savchenko ND (2001) J Optoel Adv Mater 3(2):491
28. Boolchand P, Georgiev DG, Iovu MS (2005) Chalcogenide Lett 2(4):27
29. Kumar P, Thangaraj R (2006) J Noncryst Solids 352:2288
30. Nemec P, Jedelsky J, Frumar M, Štábl M, Vlček M (2004) J Phys Chem Solids 65(7):1253
31. Yannopoulos SN, Trunov ML (2009) Phys Status Solidi B 246(8):1773
32. Igo T, Noguchi Y, Nagai H (1974) Appl Phys Lett 25:193
33. Johnson WL (1999) MRS Bull 24:42
34. Osipyan YA, Savchenko IB (1968) Pis'ma Zh Eksp Teor Fiz 7:130
35. Boyarskaya YS (1986) Physics of microindentitation processes. Stiinta, Kishinev, in Russian
36. Borisova ZU (1983) Chalcogenide semiconducting glasses. Khimiya, St. Petersburg, in Russian
37. Borisova ZU (1981) Glassy semiconductors. Plenum Press, New York, pp 215–220, and references therein

Chapter 12
Screening of Zeolitic Imidazolate Frameworks for Preconcentration of Hazardous Chemicals

Ibrahim Inanc and Ozgur Yazaydin

Abstract Zeolitic Imidazolate Frameworks (ZIFs) are porous materials which are known for their exceptional chemical/thermal stability and mostly hydrophobic character. These properties make them promising materials for use in the capture and/or detection of hazardous chemicals under humid environments. Henry's coefficient can be used in order to assess the affinity between a molecule and an adsorbent material. In this study, we used molecular simulations to find a suitable ZIF structure by a quick and easy screening method. For this screening method, the Henry's coefficients of one explosive (nitromethane), six toxic chemicals (hydrogen disulfide, sulfur dioxide, nitrogen dioxide, carbon monoxide, ethylene oxide, benzene), and three warfare agents (sarin, sulfur mustard, phosgene oxime) in pre-selected ZIFs according to their pore aperture size were computed. In addition, average loading values for the hazardous molecules under five different relative humidity conditions were obtained with GCMC simulations in ZIFs which gave the two highest Henry's coefficients with respect to the Henry coefficient of water. ZIF-1 and ZIF-68 were found to be the most promising materials for the majority of the hazardous chemicals considered in this study, with several orders of magnitude predicted preconcentration gains.

12.1 Introduction

Protection against readily available nerve agents, explosives, as well as environmental pollutants are some of the most significant issues mankind is facing worldwide. Therefore, identification of new, efficient and cheap materials

I. Inanc
Department of Chemical Engineering, University College London, Torrington Place, London WC1E7JE, UK

Department of Materials Science and Engineering, Ondokuz Mayis University, Kurupelit 55139, Samsun, Turkey

O. Yazaydin (✉)
Department of Chemical Engineering, University College London, Torrington Place, London WC1E7JE, UK
e-mail: ozgur.yazaydin@ucl.ac.uk

T.A. Camesano (ed.), *Nanotechnology to Aid Chemical and Biological Defense*, NATO Science for Peace and Security Series A: Chemistry and Biology, DOI 10.1007/978-94-017-7218-1_12

to detect, capture and remove highly toxic molecules by physical and/or chemical adsorption has attracted much attention. In this regard, screening of candidate materials for efficient capturing of highly toxic agents is of major interest for several applications, including military and industrial [1]. However, due to the broad range of toxic agents, environmental conditions and existing host materials, experimental testing of different combinations requires an enormous effort. From this perspective, computer simulations can play an important role for large-scale screening of materials in order to identify the most promising materials for further experimental investigation [2, 3]. Moreover, by gaining fundamental insights from the simulations at the microscopic level, new materials can be developed or existing materials can be modified with postsynthetic modification strategies [4].

Recently, metal-organic frameworks (MOFs) have emerged as promising materials for several applications, including the sensing and capture of toxic chemicals [5]. MOFs are nanoporous crystalline materials which are formed by the self-assembly of metal ions and organic ligands. Through coordination bonding, organic ligands bridge metal ions to form well-defined pores and high surface areas. A unique future of MOFs is the prospect of choosing different metal ions and organic linkers in order to tune the surface area, pore size and chemical functionality of these materials. However, this also means that a vast number of MOF materials can be synthesized. Indeed, more than 20,000 three dimensional MOF structures [6] have been synthesized and deposited to the Cambridge Structural Database (CSD) [7] in recent years. A particularly interesting set of materials in the MOF family is the zeolitic imidazolate frameworks (ZIFs). ZIFs are composed of tetrahedrally-coordinated transition metal ions connected by various imidazolate linkers. As their name suggest, they are topologically isomorphic with zeolites [8, 9]. ZIFs are well-known for their exceptional chemical and thermal stability and there are near 200 ZIF structures reported in the literature [10, 11]. In addition to traditional applications in gas storage and separation, ZIFs have been explored for the capture of nerve gas and mustard gas analogues, adsorption and decomposition of sarin, and detection of highly explosives and aromatic compounds [10, 12]. Harmful or toxic gases/vapors such as H_2S, SO_2, NH_3, CO, benzene have also been studied over various ZIFs [5, 10, 13].

In this work we demonstrate a molecular simulation based screening of ZIFs for preconcentration of hazardous chemicals for sensing applications. In a typical chemical sensor, the sensing mechanism depends strongly on analyte concentration. However, improving the sensitivity of a chemical sensor is often a daunting task. Instead one can utilize a preconcentrator in order to extend the detection limit of a sensor by several orders of magnitude. One other issue that needs to be addresses is the discrimination ability of sensor devices since they may be incapacitated by the higher levels of uninteresting species in the background matrix (e.g. humidity). Thus our screening strategy takes the effect of the humidity in to consideration for a more realistic outcome.

12.2 Methods

Screening strategy used in this work is rather a practical approach to gain qualitative comparison and selection purpose. In this section we give a summary of the strategy and then go in to details.

We first identified the hazardous molecules of interest and chose ten ZIF structures from the literature in order to screen them based on their preconcentration efficiency. Then we did an initial screening by computing the Henry's coefficients of the hazardous molecules and water in the selected ZIFs. Henry's coefficient is a practical way of expressing the affinity between an adsorbate molecule and the adsorbent material. Because hazardous chemicals need to be detected at extremely low concentrations, it can be safely assumed that their adsorption take place within the Henry's regime, e.g. the initial linear part of an adsorption isotherm. Then for each hazardous chemical, we identified the two most promising ZIF structures based on comparing the difference between the Henry's coefficients of the hazardous chemical and water. That is, our criterion was not to choose the ZIFs with the highest affinity for a particular hazardous molecule, but to choose the ones which showed high affinity with respect to their affinity for water. Finally, to realize the full effect of water presence on the preconcentration performance of the shortlisted ZIFs, adsorption isotherms were obtained by doing Monte Carlo simulations in the grand canonical ensemble (GCMC) for mixtures of hazardous molecule at their detection limit (see the next section) and water at five different levels of relative humidity (RH), 0 %, 25 %, 50 %, 75 % and 100 %.

12.2.1 Hazardous Molecules and ZIF Structures

In this study, we considered one explosive (nitromethane) and six toxic chemicals (hydrogen disulfide (H_2S), sulfur dioxide (SO_2), nitrogen dioxide (NO_2), carbon monoxide (CO), ethylene oxide (ETO), benzene) and three warfare agents (sarin, sulfur mustard, phosgene oxime). The detection limits for these molecules were set for 10 min exposure at their AEGL-3 (Acute Exposure Guideline Levels 3) values. AEGL-3 is the airborne concentration (expressed as ppm or mg/m^3) of a substance above which it is predicted that the general population, including susceptible individuals, could experience life-threatening health effects or death [14]. AEGL-3 values of the hazardous chemicals used in the simulations are given in the Table 12.1 with the exception of nitromethane for which highest Protective Action Criteria (PAC-3) value was used.

Next, we identified ten ZIFs with pore apertures larger than 5 Å. This was to make sure the hazardous chemicals can diffuse through the pores of the ZIFs selected for screening. Structural and chemical properties of the ZIFs considered

Table 12.1 Acute exposure guideline levels and references of force fields used for the hazardous chemicals

Molecule name/formula	AEGL-3–10 min (ppm)	Force field
Benzene	9,700	Ref. [22]
CO	1,700	Ref. [21]
ETO	360	Ref. [18]
H_2S	76	Ref. [19]
NH_3	2,700	Ref. [24]
Nitromethane	1,000 (PAC-3)	Ref. [16]
NO_2	34	Ref. [15]
Phosgene oxime	0.77	Ref. [16]
Sarin	0.06	Ref. [23]
SO_2	30	Ref. [17]
Sulfur mustard	0.6	Ref. [16]

Table 12.2 Chemical and structural details of the ZIFs used in the simulations [11]

ZIF name	ZIF formula[a]	Pore aperture (Å)	Pore diameter (Å)
HIFVOI	$Zn(Im)_2$	6.6	9.6
HIFVUO	$Zn(Im)_2$	5.2	8.6
VEJYIT-01	$Zn(Im)_2$	5.4	5.7
ZIF-1	$Zn(Im)_2$	6.3	6.9
ZIF-2	$Zn(Im)_2$	6.4	6.9
ZIF-10	$Zn(Im)_2$	8.2	12.1
ZIF-60	$Zn_2(Im)_3(mIm)$	7.2	9.4
ZIF-68	Zn(bIm)(nIm)	7.5	10.3
ZIF-70	$Zn_{1.13}(Im)_{1.13}(nIm)_{0.87}$	13.1	15.9
ZIF-80	Zn(dcIm)(nIm)	9.8	13.2

[a]See Fig. 12.2 for the open formulas of the ligands

in this study are given in Table 12.2. The positions of the ZIF atoms were taken from experimental crystallographic data (Fig. 12.1) [7]. All the ZIFs studied in this work have Zn metal centers connected by different imidazolate-based linkers. The chemical structures of the organic linkers (Imidazolate (Im), 2-methylimidazolate (mIM), benzimidazolate (bIM), 2-nitroimidazolate (nIM), 4,5-dichloroimidazolate (dcIM)) are given in Fig. 12.2. Except nIM these ligands are all hydrophobic. Therefore, the seven ZIFs; HIFVOI, HIFVUO, VEJYIT-01, ZIF-1, ZIF-2, ZIF-10, and ZIF-60, are hydrophobic in nature because Im and/or mIM are used as bridging ligands in these ZIFs. On the other hand, ZIF-68, ZIF-70, and ZIF-80 have hydrophobic pores decorated with bIm, Im, and dcIm, respectively. However, they also have hydrophilic pores with nIM linkers. Therefore, these three ZIFs have both hydrophobic and hydrophilic channels.

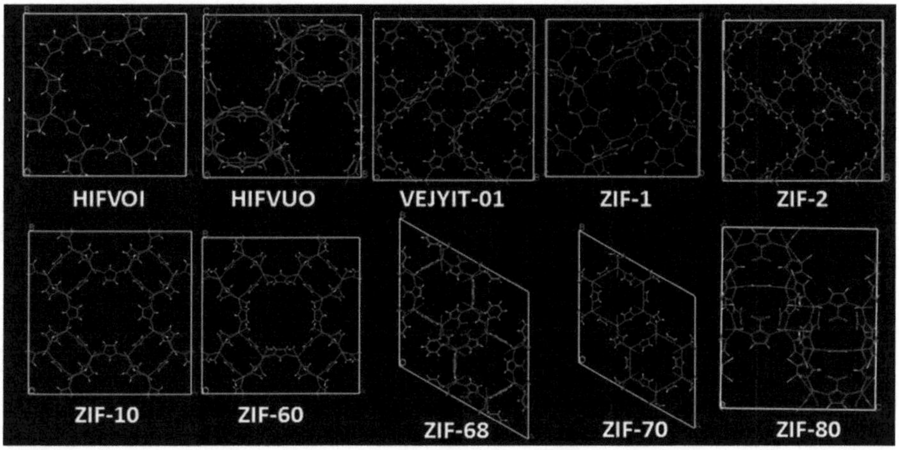

Fig. 12.1 Unit cells of ZIF structures used in the simulations (*White* (H), *gray* (C), *blue* (N), *red* (O), *green* (Cl))

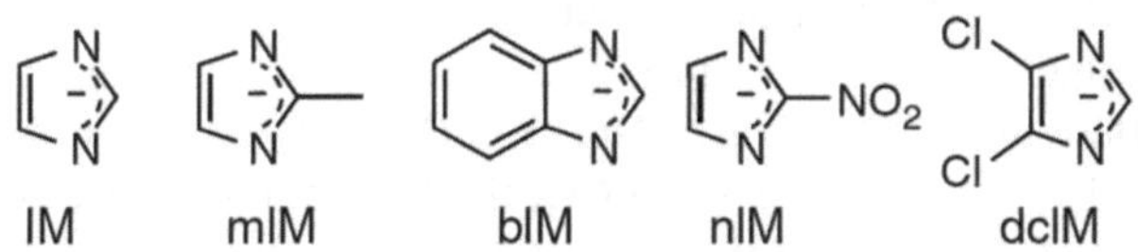

Fig. 12.2 Organic ligands used in ZIFs considered in this study (*IM* imidazolate, *mIM* 2-methylimidazolate, *bIM* benzimidazolate, *nIM* 2-nitroimidazolate, *dcIM* 4,5-dichloroimidazolate)

12.2.2 Force Field and Molecular Simulations

In our molecular simulations, the interaction energies between all gas and ZIF atoms were computed through a combination of Lennard-Jones (LJ) and Coulomb potentials

$$V_{ij} = 4\,\epsilon_{ij}\left[\left(\frac{\sigma_{ij}}{r_{ij}}\right)^{12} - \left(\frac{\sigma_{ij}}{r_{ij}}\right)^{6}\right] + \frac{q_i q_j}{4\epsilon_o r_{ij}} \tag{12.1}$$

where i and j are the interacting atoms, and r_{ij} is the distance between the atoms i and j. ε_{ij} and σ_{ij} are LJ well depth and diameter, respectively. qi and qj are the partial charges of the interacting atoms, and ε_0 is the dielectric constant. LJ parameters between different types of atoms were calculated using the Lorentz–Berthelot mixing rules with a 12 Å cut off distance for non-bonded interactions. To calculate the electrostatic interactions the Ewald summation technique was used.

Force field parameters for the hazardous chemicals were obtained from literature. Table 12.1 lists the source of these parameters [15–24]. For water the TIP5P-Ew model was used [25]. All hazardous molecules and water were treated as rigid molecules. For the MOF atoms the LJ parameters were taken from a refined version of the universal force field (UFF) [26–28]. Partial atomic charges were calculated by the EQE method proposed by Wilmer and Snurr [29].

Henry's coefficients were calculated using Widom's test particle method [30] by a 5×10^4 production run. GCMC simulations of mixtures included random insertion, deletion, translation, rotation, regrowth and identity exchange moves between different species. Simulations were first equilibrated for 10^5 cycles which was followed by a 2×10^5 cycle production run. Mol fractions of hazardous molecules and water in the mixtures were set based on the detection limits (see Table 12.1) and the partial pressure of water in air as a function of relative humidity, respectively. All simulations were performed with RASPA package [31] at 298 K. During the simulations ZIF atoms were held fixed at their crystallographic positions.

12.3 Results and Discussion

Henry's coefficients (HCs) of all hazardous chemicals and water obtained from molecular simulations are shown in Fig. 12.3a, b. For small gases, the predicted HC values follow the ETO > SO_2 > NO_2 > NH_3 > H_2S > CO order in the seven hydrophobic ZIFs (Fig. 12.3a). Except CO, the HC values are greater than the HC of water in these ZIFs, whereas the HC of CO is very close to that of predicted for water. On the other hand, in ZIF 68, 70 and 80 water has higher HCs. NH_3 and H_2S have HC values very close to water, whereas ETO and SO_2 have higher HCs compared to water in these three ZIFs.

For the large hazardous molecules (Fig. 12.3b), the predicted HCs of sarin is less than the HC of water in all ZIFs. This means, none of the ZIFs considered in this study will adsorb sarin preferentially over under the presence of humidity. Therefore, sarin was not included in the next screening step. Sulfur mustard has higher HC values than water only in ZIF-68 and ZIF-70 which indicates that only these two structures may be useful under humid conditions. Phosgene, nitromethane and benzene, on the other hand, have many orders of magnitude higher values of HC with respect to the HC of water.

In the last step of the screening method, we investigated the effect of increasing levels of humidity on the preconcentration effect of ZIFs by GCMC simulations. For each hazardous chemical, except sarin, the two ZIF structures that gave the highest difference between the Henry's coefficients of the hazardous chemical and water were considered (Table 12.3). For convenience the results are grouped in to four different sets in Figs. 12.4, 12.5, 12.6, and 12.7. In all figures, dashed lines show the AEGL-3 level of the hazardous chemicals. At %0 RH we observe that the uptakes of hazardous chemicals are well above the ambient AEGL-3 values, which indicates that all ZIFs show preconcentration effect under dry conditions. Here we define the preconcentration gain as the ratio of the concentration of the hazardous chemical in the adsorbed phase to its AEGL-3 level in air.

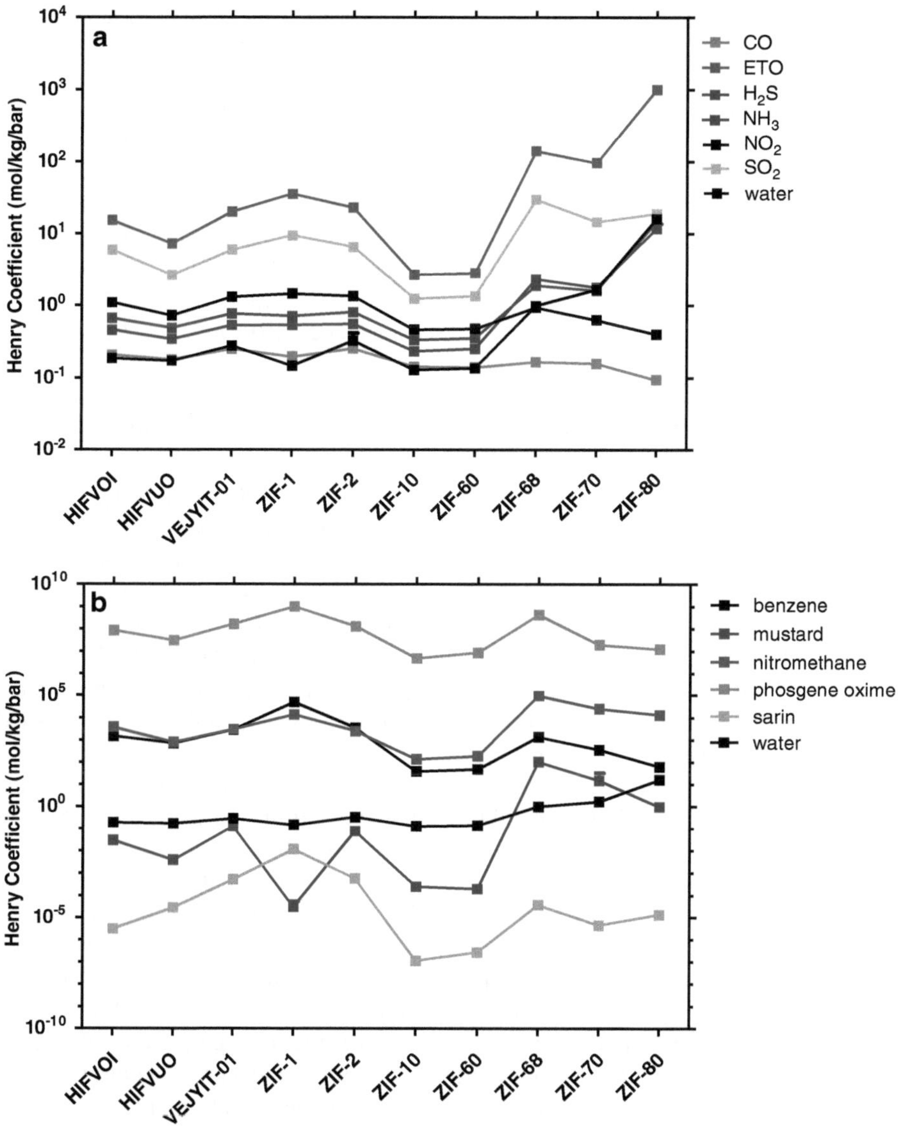

Fig. 12.3 Henry's coefficients of (**a**) small, and (**b**) large hazardous chemicals in ZIF structures in comparison with the Henry's coefficient of water

In all ZIFs water uptake increases as RH increases. Despite this, however, in general the uptakes of the hazardous molecules are not affected significantly with increasing levels of humidity. Exceptions are H_2S at 25 % RH in ZIF-1 and ZIF-68, and NH_3 and ETO only in ZIF-80. Furthermore, even at the saturated levels of

Table 12.3 ZIFs which gave the two highest Henry's coefficient with respect to water for hazardous chemicals. The ZIFs listed below were used in simulating the mixture isotherms

Molecule name	1st structure	2nd structure
Benzene	ZIF-1	ZIF-68
CO	ZIF-1	HIFVOI
ETO	ZIF-1	ZIF-80
H_2S	ZIF-1	ZIF-68
NH_3	ZIF-1	ZIF-80
Nitromethane	ZIF-1	ZIF-68
NO_2	ZIF-1	ZIF-2
Phosgene oxime	ZIF-1	ZIF-68
SO_2	ZIF-1	ZIF-68
Sulfur mustard	ZIF-70	ZIF-68

humidity (100 % RH), the adsorption levels of the hazardous molecules are much higher than their AEGL-3 levels.

Figure 12.4 shows the average loading values of CO, SO_2, and NO_2. HIFVOI and ZIF-1 provide preconcentration gains of four and five, respectively, for CO in comparison to the AEGL-3 level of CO; they are not affected by increasing %RH. For SO_2, ZIF-1 and ZIF-68 are predicted to give preconcentration gains of roughly 260 and 660, respectively, at all %RHs, except at 25 % RH where the uptake of SO_2 significantly increases in both ZIFs and decrease back at 50 % RH. This can be explained by the fact that at low loadings, water molecules provide adsorption sites for SO_2 molecules. However, as water uptake increases SO_2 molecules are pushed out of the pores. We observed this phenomenon previously between SO_2 and water, and SO_2 and CO_2 in other MOFs [27, 32]. For NO_2 predicted preconcentration gains are approximately 35 in ZIF-1 and 24 in for ZIF-2. Similar to CO, NO_2 uptake in both ZIFs are not affected significantly by increasing %RH.

Figure 12.5 shows the results for NH_3, ETO, and H_2S that have more hydrogen/polarity in their structure. ZIF-1 provides a preconcentration gain of approximately 970, 26, and 16 for ETO, NH_3, and H_2S, respectively, irrespective of the %RH. On the other hand, in ZIF-80 preconcentration gain for NH_3 is 370 and for ETO 15300. However, as RH increases preconcentration gains in ZIF-80 for NH_3 and ETO decrease by about one third. In contrast, the average loading of H_2S in ZIF-68 does not change significantly with the humidity, and a preconcentration gain of about 53 is predicted with respect to the AEGL-3 concentration of H_2S.

In Fig. 12.6 the average loading values for nitromethane, phosgene oxime and benzene are shown under increasing %RH. For all these three molecules ZIF-1 and ZIF-68 gave the highest difference between their HCs and the HCs of water. ZIF-1 and ZIF-68 provide preconcentration gains of about 5,154 and 7,216 for benzene, 17,000 and 12,000 for nitromethane, and 8×10^6 and 12×10^6 for phosgene oxime, respectively. In both ZIF-1 and ZIF-68 the predicted preconcentration gains are not affected by increasing humidity levels. The main reason for the very high preconcentration gains in these ZIFs is apparent from the very large HC values computed for these molecules, which indicate strong affinity between the host and guest molecules.

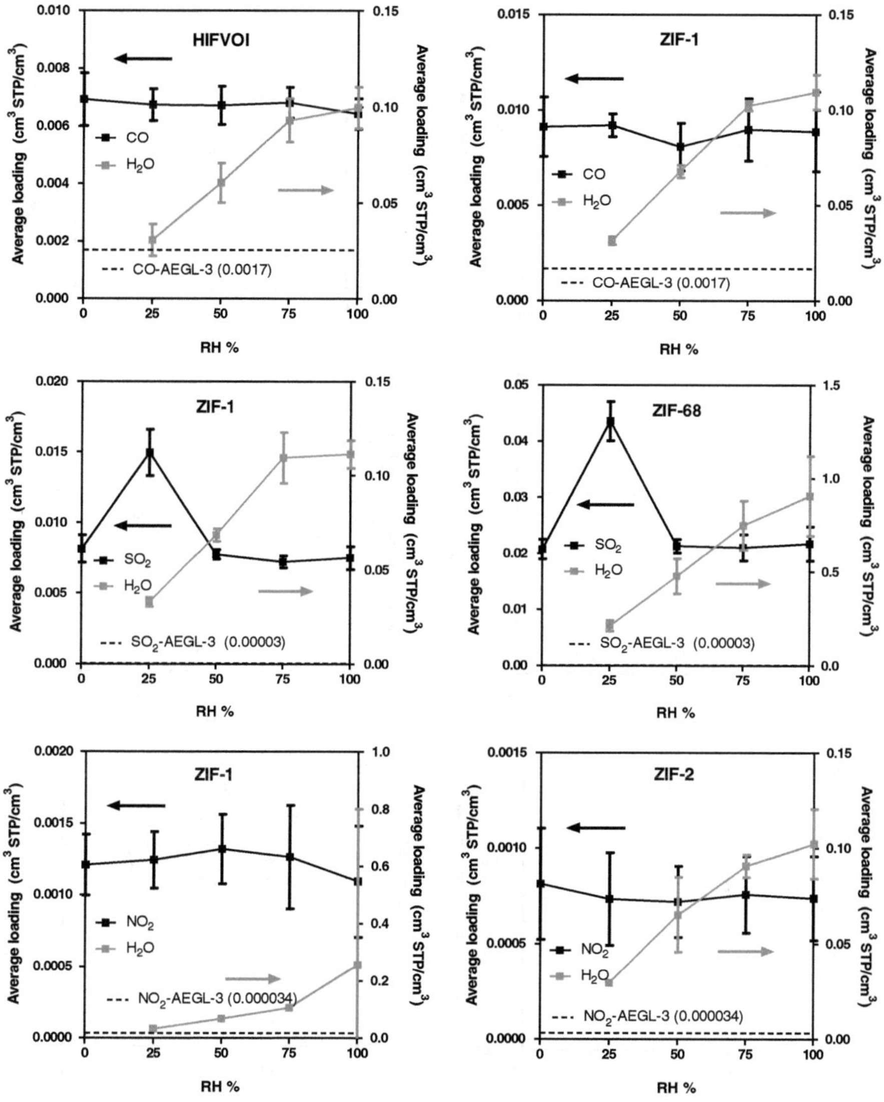

Fig. 12.4 Effect of humidity on the average loading values of the hazardous chemicals; CO, SO_2 and NO_2, in ZIFs which gave the two highest Henry's coefficient with respect to water. *Dashed lines* represent the AEGL-3–10 min levels for the hazardous molecules

Finally, Fig. 12.7 shows the results for sulfur mustard. Under dry conditions ZIF-68 and ZIF-70 show some preconcentration effect; however, with the introduction of the humidity the effect is lost. This means both ZIFs will not work as a preconcentrator for sulfur mustard sensing.

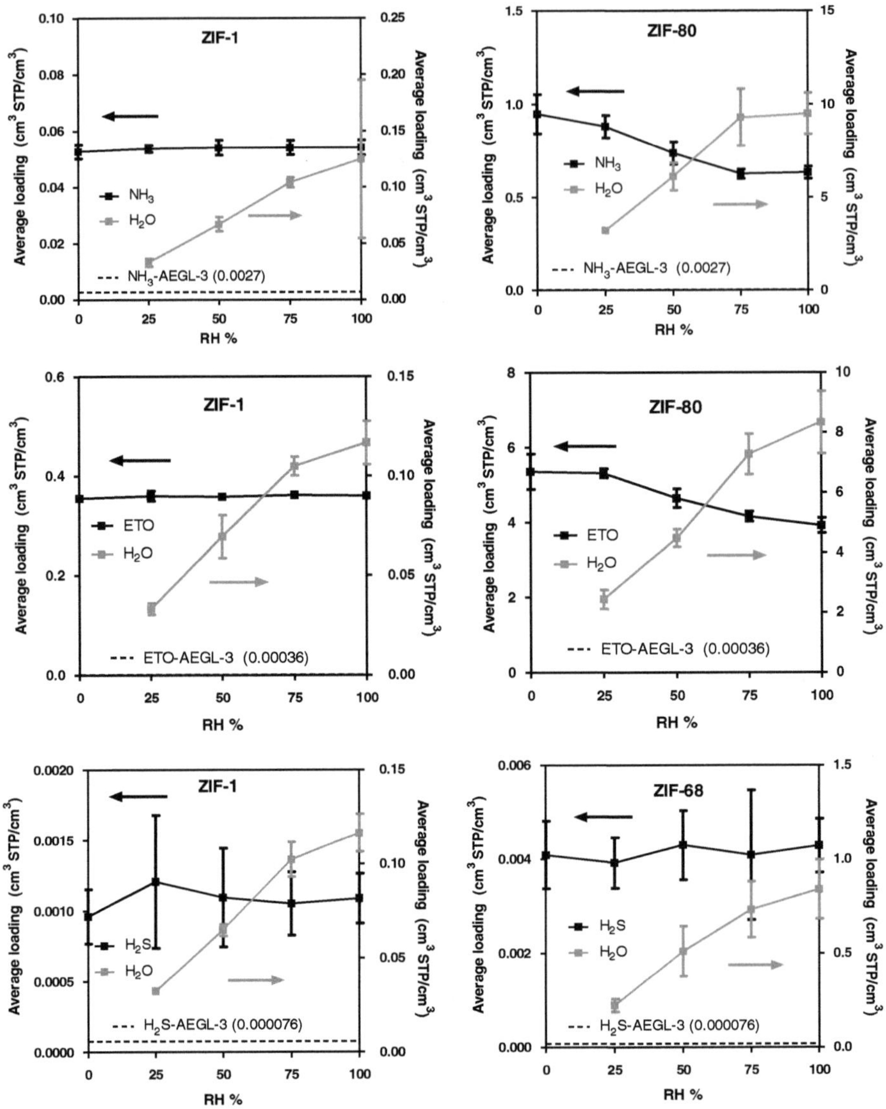

Fig. 12.5 Effect of humidity on the average loading values of the hazardous chemicals; NH_3, ETO and H_2S, in ZIFs which gave the two highest Henry's coefficient with respect to water. *Dashed lines* represent the AEGL-3–10 min levels for the hazardous molecules

12.4 Conclusions

Ten hazardous chemicals and ten ZIFs were screened with molecular simulations. Two ZIF structures, ZIF-1 and ZIF-68 emerged as promising structures to provide preconcentration gains for the sensing of the hazardous chemicals considered in this

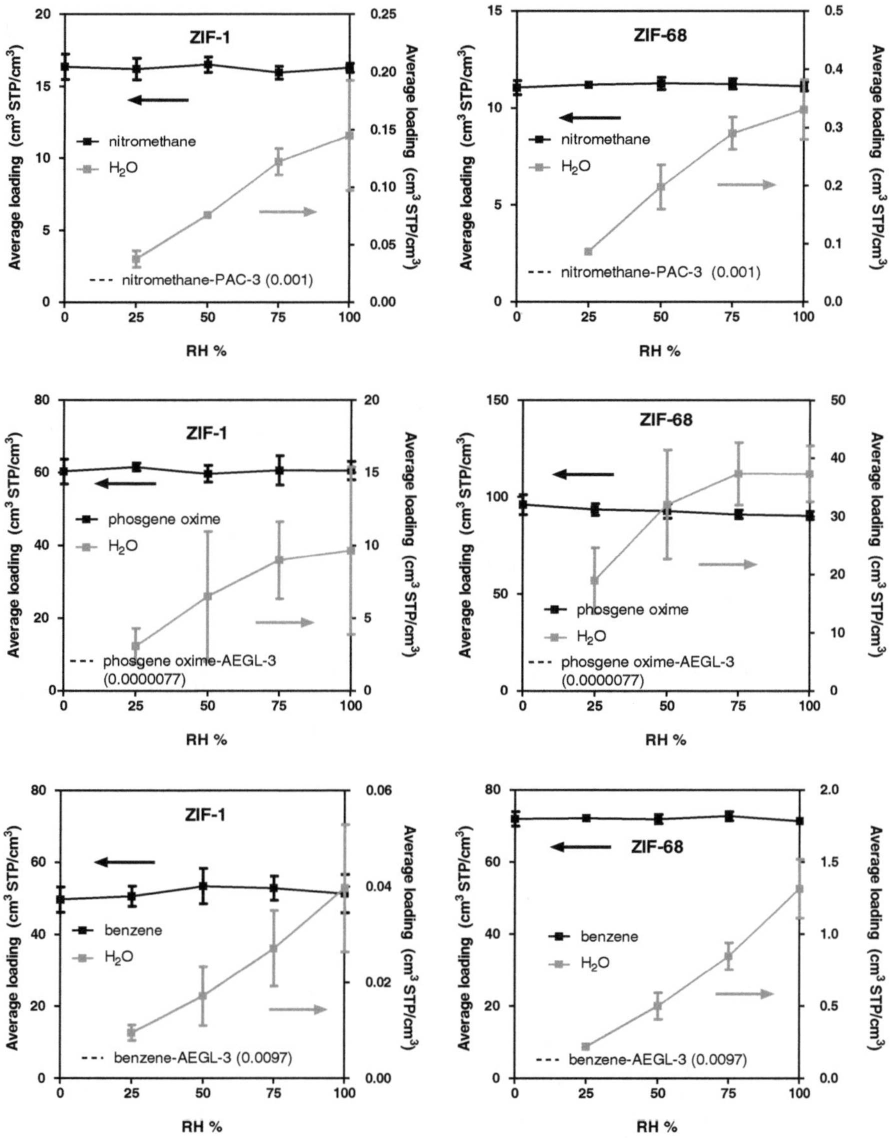

Fig. 12.6 Effect of humidity on the average loading values of the hazardous chemicals; nitromethane, phosgene oxime and benzene, in ZIFs which gave the two highest Henry's coefficient with respect to water. *Dashed lines* represent the AEGL-3–10 min levels for the hazardous molecules

study, except sarin and sulfur mustard. In addition, the predicted preconcentration gains were not significantly affected with increasing levels of relative humidity. For large hazardous chemicals; benzene, nitromethane and phosgene oxime,

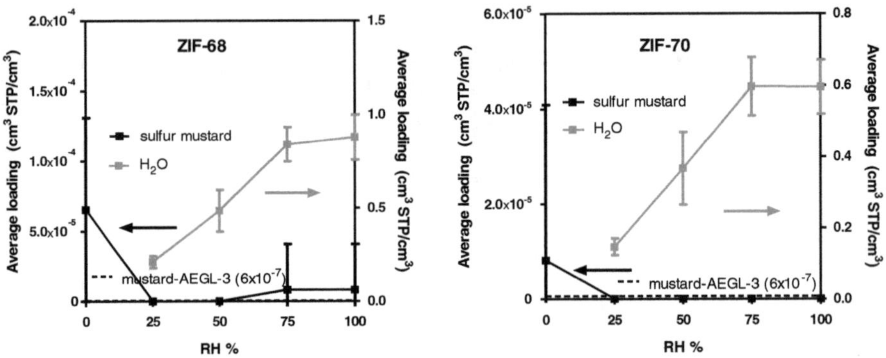

Fig. 12.7 Effect of humidity on the average loading values of the sulfur mustard in ZIFs which gave the two highest Henry's coefficient with respect to water. *Dashed lines* represent the AEGL-3–10 min levels for the hazardous molecules

preconcentration gains ranged between three to seven orders of magnitude with respect to their AEGL-3–10 min levels, whereas for small hazardous chemicals, this effect ranged from one (CO, NO_2, H_2S) to three (SO_2, ETO) orders of magnitude.

Acknowledgments Inanc has been supported by the TUBITAK (The Scientific and Technological Research Council of Turkey) under the Program No. 2219 (International Post-doctoral Research Fellowship Programme) for this research.

References

1. Greathouse JA et al (2010) Computational screening of metal-organic frameworks for large-molecule chemical sensing. Phys Chem Chem Phys 12(39):12621–12629
2. Sarkisov L (2012) Toward rational design of metal-organic frameworks for sensing applications: efficient calculation of adsorption characteristics in zero loading regime. J Phys Chem C 116(4):3025–3033
3. Colon YJ, Snurr RQ (2014) High-throughput computational screening of metal-organic frameworks. Chem Soc Rev 43(16):5735–5749
4. Deria P et al (2014) Beyond post-synthesis modification: evolution of metal-organic frameworks via building block replacement. Chem Soc Rev 43(16):5896–5912
5. Khan NA, Hasan Z, Jhung SH (2013) Adsorptive removal of hazardous materials using metal-organic frameworks (MOFs): a review. J Hazard Mater 244:444–456
6. Chung YG et al (2014) Computation-ready, experimental metal-organic frameworks: a tool to enable high-throughput screening of nanoporous crystals. Chem Mater 2:766–774
7. Allen FH (2002) The Cambridge structural database: a quarter of a million crystal structures and rising. Acta Crystallogr B Struct Sci 58:380–388
8. Banerjee R et al (2008) High-throughput synthesis of zeolitic imidazolate frameworks and application to CO2 capture. Science 319(5865):939–943
9. Park KS et al (2006) Exceptional chemical and thermal stability of zeolitic imidazolate frameworks. Proc Natl Acad Sci U S A 103(27):10186–10191

10. DeCoste JB, Peterson GW (2014) Metal-organic frameworks for air purification of toxic chemicals. Chem Rev 114(11):5695–5727
11. Phan A et al (2010) Synthesis, structure, and carbon dioxide capture properties of zeolitic imidazolate frameworks. Acc Chem Res 43(1):58–67
12. Barea E, Montoro C, Navarro JAR (2014) Toxic gas removal – metal-organic frameworks for the capture and degradation of toxic gases and vapours. Chem Soc Rev 43(16):5419–5430
13. Song XD et al (2014) Investigation of SO2 gas adsorption in metal-organic frameworks by molecular simulation. Inorg Chem Commun 46:277–281
14. http://www.epa.gov/oppt/aegl/pubs/define.htm
15. Bourasseau E et al (2008) Thermodynamic behavior of the CO2 + NO2/N2O4 mixture: a Monte Carlo simulation study. J Phys Chem B 112(49):15783–15792
16. Dauberosguthorpe P et al (1988) Structure and energetics of ligand-binding to proteins – Escherichia-coli dihydrofolate reductase trimethoprim, a drug-receptor system. Proteins Struct Funct Genet 4(1):31–47
17. Ketko MH, Kamath G, Potoff JJ (2011) Development of an optimized intermolecular potential for sulfur dioxide. J Phys Chem B 115(17):4949–4954
18. Ketko MH et al (2008) Development of the TraPPE-UA force field for ethylene oxide. Fluid Phase Equilib 274(1–2):44–49
19. Kristof T, Liszi J (1997) Effective intermolecular potential for fluid hydrogen sulfide. J Phys Chem B 101(28):5480–5483
20. Martin MG, Siepmann JI (1998) Transferable potentials for phase equilibria. 1. United-atom description of n-alkanes. J Phys Chem B 102(14):2569–2577
21. Martin-Calvo A, Lahoz-Martin FD, Calero S (2012) Understanding carbon monoxide capture using metal organic frameworks. J Phys Chem C 116(11):6655–6663
22. Rai N, Siepmann JI (2007) Transferable potentials for phase equilibria. 9. Explicit hydrogen description of benzene and five-membered and six-membered heterocyclic aromatic compounds. J Phys Chem B 111(36):10790–10799
23. Sokkalingam N et al (2009) Extension of the transferable potentials for phase equilibria force field to dimethylmethyl phosphonate, sarin, and soman. J Phys Chem B 113(30):10292–10297
24. Zhang L, Siepmann JI (2010) Development of the trappe force field for ammonia. Collect Czechoslov Chem Commun 75(5):577–591
25. Mahoney MW, Jorgensen WL (2000) A five-site model for liquid water and the reproduction of the density anomaly by rigid, nonpolarizable potential functions. J Chem Phys 112(20):8910–8922
26. Rappe AK et al (1992) Uff, a full periodic-table force-field for molecular mechanics and molecular-dynamics simulations. J Am Chem Soc 114(25):10024–10035
27. Ding LF, Yazaydin AO (2013) The effect of SO2 on CO2 capture in zeolitic imidazolate frameworks. Phys Chem Chem Phys 15(28):11856–11861
28. Liu B, Smit B (2010) Molecular simulation studies of separation of CO2/N-2, CO2/CH4, and CH4/N-2 by ZIFs. J Phys Chem C 114(18):8515–8522
29. Wilmer CE, Kim KC, Snurr RQ (2012) An extended charge equilibration method. J Phys Chem Lett 3(17):2506–2511
30. Vlugt TJH et al (2008) Computing the heat of adsorption using molecular simulations: the effect of strong Coulombic interactions. J Chem Theory Comput 4(7):1107–1118
31. Dubbeldam D, Calero S, Ellis DE, Snurr RQ (2008) RASPA. Northwestern University, Evanston
32. Ding LF, Yazaydin AO (2012) How well do metal-organic frameworks tolerate flue gas impurities? J Phys Chem C 116(43):22987–22991

Chapter 13
Quasi-Fulleranes and Fulleranes as Main Products of Fullerenizasion of Molecules of Benzene, Toluene and Pyridine

Oleksii Kharlamov, Marina Bondarenko, Ganna Kharlamova, and Veniamin Fomenko

Abstract Intensive 25 year attempts of researchers with use of superhigh pressures and temperatures to receive maximum saturated by hydrogen fullerane ($C_{60}H_{60}$) by means of heterophases reactions of hydrogenation of fullerite (or fullerene) have appeared completely unsuccessful. This fact from the chemical point of view was quite predicted: in heterophase reactions limiting always is the stage of diffusion of one of reagents (in particular, hydrogen) through a layer of the resultant product (in particular, fullerane) therefore the particle of a final product (fullerane) is always non-uniform on composition. Moreover, defullerenization of molecules C_{60} and $C_{60}H_x$ at temperatures higher 400 °C is observed as at hydrogenation of first, and dehydrogenation of second. Herein is described a new (author method) method of pyrolysis as process of fullerenization (transformation of organic molecules in the closed molecules of carbon and their hydrides) of molecules of benzene, toluene and pyridine, which has allowed to obtain simultaneously and carbon molecules, and them hydrides. Equiatomic composition fullerane $C_{60}H_{60}$, quasi-fulleranes ($C_nH_{n\text{-}6}$–C_nH_{n-2} ($n = 20$–46)) and quasi-fullerenes (C_{48}, C_{40}, C_{42}) as well as small carbon molecules (C_3–C_{11}) are detected only in mass spectra of products of fullerenization of organic molecules. In products of fullerenization of pyridine molecules (C_5H_5N) are detected new heteroatomic molecules such as hydrogenated and hydroxylated azafullerenes $(C_{35}N_5)H_9$, $(C_{45}N_5)(OH)_3H_{14}$ and $(C_{49}N_{11})(OH)_5H_{18}$ as well as homoatomic molecules C_{60}, C_{48} and C_3–C_{18}. This method allows obtain fulleranes and quasi-fulleranes in gramme quantities.

O. Kharlamov (✉) • M. Bondarenko
Frantsevich Institute for Problems of Materials Science of NASU, Krzhyzhanovsky St. 3, 03142 Kiev, Ukraine
e-mail: dep73@ipms.kiev.ua

G. Kharlamova
Taras Shevchenko National University of Kyiv, 64 Volodymyrska str., 03001 Kyiv, Ukraine

V. Fomenko
National University of Food Technologies, 68 Volodymyrska str., 03001 Kiev, Ukraine

T.A. Camesano (ed.), *Nanotechnology to Aid Chemical and Biological Defense*, NATO Science for Peace and Security Series A: Chemistry and Biology,
DOI 10.1007/978-94-017-7218-1_13

13.1 Introduction

Derivatives of fullerene such as exohedral and endohedral fullerenes, exohedral adducts and heterofullerenes (azafullerenes, azabora(C_{60})fullerene) are increasingly used as catalysts, sorbents, superconductors, lubricants, polymeric composites, antioxidants and nanosensors. Fulleranes (or hydrides of fullerenes) as the first fullerene derivatives have attracted a special attention as a supposedly perspective and efficient accumulators of hydrogen. Optimistically it was supposed that the electron deficient molecules of fullerenes (E for $C_{60} = -0{,}44$ v and for $C_{70} = -0{,}41$ v) can easily be hydrogenated to form an equiatomic composition of molecules of fulleranes $C_{60}H_{60}$ and $C_{70}H_{70}$. Also it was supposed, that at fulleranes dehydrogenation and the destruction of bonds C–H close on energy bonds C–C of fullerene cage to collapse will not be. However, the first experiments [1] were showed that C_{60} molecule is surprising not only on the structure: its hydrogenation is fulfilled easily either on 30 % or on 60 % till forming of fulleranes $C_{60}H_{18}$ and $C_{60}H_{36}$, respectively, which, consequently, are most stable. It is well known, that the molecules with the conjugated double bonds have the increased stability. It is possible, that partially hydrogenated molecules fullerene $C_{60}H_{18}$ and $C_{60}H_{36}$ have new (fullerene) steady conjugation of stayed double bonds. Numerous attempts to obtain fulleranes with more high content of hydrogen were undertaken further. However, an increased (and greatest) degree (~73 %) of hydrogenation of the fullerene was only achieved in two stage methods or in the methods, in which the superhigh pressure and catalysts were used. Therefore the search of new approaches for development more efficient methods of synthesis of hydrogenated carbon molecules, namely fulleranes ([1–3], [5–11]) and quasi-fulleranes [4, 12, 13], is extremely urgent.

In the present report is discussed new (distinct from already of known methods (continuous flowing and flash)) method of pyrolysis as process of fullerenization of organic molecules, in particular aromatic hydrocarbons, pyridine and ethanol [4, 12, 13]. The process fullerenization is reactions of transformation of molecules of precursor and fragments of them destruction in the closed carbon molecules, them hydrogenated and hetero derivatives (It should to remind, that in the beginning of the last century the extremely important industrial process of aromatization of molecules saturated hydrocarbons (for example, hexane) in aromatic molecules, in particular benzene as well as pentadiene, was developed). Distinctive feature of an author's method of pyrolysis (or process of fullerenization of organic molecules) is the opportunity at rather low temperatures (~>700 °C) except for known carbon molecules, for example C_{60}, simultaneously to generate new (earlier detected only in carbon plasma clusters) carbon molecules (quasi-fullerenes and small carbon molecules) as well as them hydrogenated derivatives. Is remarkable, that only at fullerenization process the formation in gramme quantities of fulleranes (as well as quasi-fulleranes) is carry out in one stage without use of a stage of preliminary synthesis of carbon molecules. Moreover, only in process of fullerenization owing to homogeneous radical reactions (instead of heterogeneous reactions at hydrogenation of fullerite) it is possible to receive completely saturated hydrogenated carbon molecules, in particular eguiatomic fullerane $C_{60}H_{60}$ [12].

Herein the special attention also will be given to the study by mass spectrometric analysis of the gaseous products (and primarily hydrogen) which are formed at thermal decomposition in range temperature 50–750 °C of mixture of condensed hydrogenated molecules (fulleranes and quasi-fulleranes as well as hydrogenated small carbon molecules C_3–C_{15}) of synthesized at fullerenization of molecules of benzene, toluene and pyridine [4, 11, 12].

13.2 Main Known Methods for the Synthesis of Fullerenes and Fulleranes

Before discovery by Kroto [14] of a unique carbon molecule C_{60} Rofling [15] by means of mass spectrometric method was detected more 50 even and odd carbon clusters (Fig. 13.1a) in carbon plasma generated by a laser irradiation of graphite. However Kroto [14] has created such technological conditions, at which extremely active clusters of the small sizes in such generated carbon plasma coalesced mainly in larger and steady cluster, which has appeared of the Nobel advantage a molecule fullerene C_{60}. Later Kratschmer [19] has developed the technique of reception of fullerene C_{60} in microquantity from carbon plasma generated by an arc discharge (~3500 °C) method. Except for fullerene C_{60} from soot formed by an arc method (fullurene soot) it was possible to allocate of its greater size homologies C_{70}, C_{76}, C_{78}, C_{84}, in which structure as well as in C_{60} are absent adjacent pentagons.

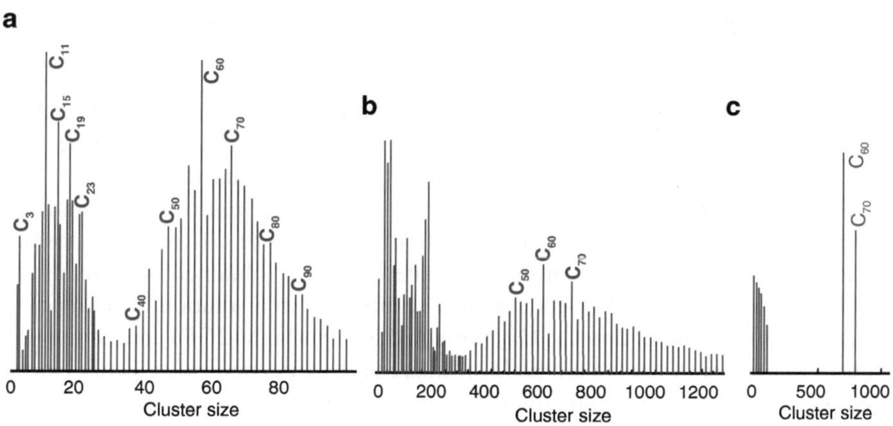

Fig. 13.1 Schematic illustration of mass spectra of: obtained by Rohlfing [15] carbon plasma generated by laser irradiation of graphite (**a**); sooting flame incomplete combustion of hydrocarbons presented in [16, 17] (**b**); fullerene soot obtained laser ablation of graphite [18] (**c**)

Along with super high-temperature (~3500 °C) methods of sublimation of carbon (laser ablation and arc discharge) for obtaining of carbon molecules was de-veloped also high-temperature (~1800 °C) method of burning of hydrocarbons, in particular, benzene in a smoky flame [20]. At burning of benzene is also generated, as it is visible from mass spectrum of its gas phase (Fig. 13.1b [16, 17]) wide range of carbon clusters of the different sizes. Just these methods (arc discharge and incomplete combustion of hydrocarbons) are nowadays basic for synthesis of fullerene C_{60} and its homologies (C_{70}, C_{76}, C_{78}, C_{84} and more) [21]. However, most part of clusters of the smaller sizes to isolate from carbon plasma or a smoky flame as molecules has appeared while impossible. So, fullerene soot obtained by laser ablation of graphite contains only fullerenes C_{60} and C_{70}, but does not contain carbon molecules with number of atoms of carbon less than 60 (Fig. 13.1c) [18].

Thus, majority fullerenes of the smaller size C_n ($n<60$) (quasi-fullerenes), in which isolated pentagons absent, and the small molecules of carbon with even and odd number of atoms of carbon C_n ($n<20$) by means of such highly temperature methods are not for the present synthesized. As the evaporation of carbon is carried out in an inert atmosphere which is not containing hydrogen, the absence of hydrogenated carbon molecules in fullerene soot of these methods is quite explained. Probably, therefore Haufler has applied a traditional method for obtaining of fullerene hydride. In 1990 year Haufler [1] has fulfilled the hydrogenation of fullerene C_{60} on Birch and has obtained the first fullerane $C_{60}H_{36}$, which has become first of derivative of fullerene. Later have begun to synthesis and others derivative of fullerene, such as exohedral and endohedral fullerenes, exohedral adducts and heterofullerenes. However fulleranes have attracted the special attention of the experts in the field of hydrogen power in quality ostensibly of perspective and effective accumulators of hydrogen. The synthesis of fullerane of equiatomic composition $C_{60}H_{60}$ containing 7.7 % of hydrogen would allow to obtain a perspective material for hydrogen power. (Though it is necessary to note, that it is enough easily obtained dodecahedrane $C_{20}H_{20}$ [22] contains also 7.7 % of hydrogen).

All known techniques of synthesis of fulleranes, in essence, are based on one method, namely, the hydrogenation previously synthesized fullerene in a solution or fullerite by various hydrogencontaining reagents (Fig. 13.2) ([1–3], 2015; [5–11]). Depending on hydrogenated reagent and reactionary conditions (temperature, pressure, catalyst, concentration) fulleranes of different composition were obtained: from minimal hydrogenated $C_{60}H_2$ (obtained selectively by hydroboration [23] or hydrozirconation [24] fullerene in a solution of toluene) up to fulleranes of composition $C_{60}H_{52}$ (obtained at superhigh (2GPa) pressure [10]). However majority of known methods of fullerenes hydrogenation (by Birch-Huckel method, Zn-concHCl system, hydrogen radical-induced hydrogenation, transfer hydrogenation by 9,10-dihydroanthracene etc) is resulted to obtaining steadiest fullerane of composition $C_{60}H_{36}$ [1–3, 5–11].

Attempts to obtain hydrides of fullerenes with the large contents of hydrogen were indispensable finished by essential complication of process of hydrogenation. So, highly hydrogenated fullerane $C_{60}H_{44}$ (with an impurity of minor fractions

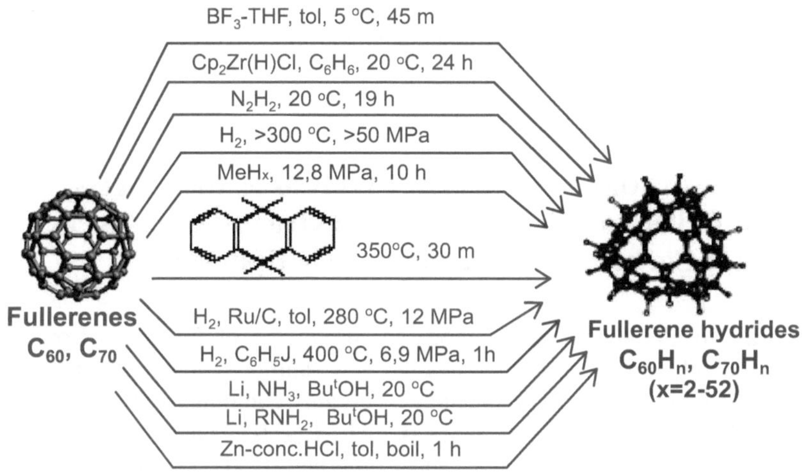

Fig. 13.2 A schematic view of known methods of synthesis of fulleranes from fullerenes (or fullerite)

of composition $C_{60}H_{46}$-$C_{60}H_{48}$) [7, 8] was obtained only with use of a two-stage method: synthesized on a Birch-Huckel method fullerane $C_{60}H_{36}$ at the second stage in addition hydrogenate on a method of Benkeser (by means of reduction it by lithium in ethylene diamine at presence of tert-buthanol (*t*-BuOH)). A mixture of fulleranes of composition $C_{60}H_{48}$-$C_{60}H_{50}$ synthesize by a method liquid-phase (in toluene solution) catalytic hydrogenation C_{60} by hydrogen at pressure 12 MPa and at presence of metal-carbon (5 % Ru/C) catalyst [9], that, naturally, reduces an output of a fullerane. According to [10] the mixture of hydrides $C_{60}H_{44}$-$C_{60}H_{52}$ of fullerene C_{60} can be prepared during 30–90 min in conditions of direct hydrogenation of fullerite at ultra-high (2 GPa) pressure and increased (>450 °C) temperatures.

The maximal contents of hydrogen in hydrogenated fullerite, answering, ostensibly, fulleranes $C_{60}H_{60}$ [2], was achieved only in conditions by its long (more than 10 h) heat treatment at temperatures above 450 °C and superhigh pressure (1–12 MPa). However, probably, that hydrogenation of fullerite at so high pressure (9.8 MPa) and temperatures (>600 °C) results in destruction of molecules of synthesized fullerane with formation of smaller fragments $C_{59}H_x$, $C_{58}H_x$, $C_{57}H_x$ etc [2, 5, 6, 10]. The fragments formed as radicals it is easy hydrogenate up to thermostable hydrocarbons. The saturation in such destroyed fragments of bonds C–C by hydrogen also creates illusion of reception of limiting saturated by hydrogen fullerane $C_{60}H_{60}$.

Thus, all developed for today techniques (Fig. 13.2) [2, 5, 6, 10, 11] of a synthesis of fulleranes are based on one process, namely, the process of hydrogenation of previously synthesized fullerene, fullerite or palladium fulleride. Fulleranes, in turn, obtain mainly at superhigh temperatures on condition of carbon sublimation.

Therefore the search of the new approaches for development of more effective methods of reception, as carbon molecules, and their hydrogenated derivatives is still urgent.

13.3 New (Authors) Method of Pyrolysis as the Process of Fullerenization Molecules of Organic Substances into Carbon Molecules and Their Hydrides

In contrast to already known methods for producing of fullerenes from carbon plasma and fulleranes from fullerenes the new (author's) method of pyrolysis as process of fullereniization of molecules of organic compounds into the molecules not only fullerene C_{60} and fulleranes ($C_{60}H_8$–$C_{60}H_{60}$ and $C_{70}H_8$–$C_{70}H_{44}$), but also quasi-fullerenes C_{40}, C_{42}, C_{48}, quasi-fulleranes (C_nH_{n-6}–C_nH_{n-2} (n = 20–46)) and small carbon molecules C_3–C_{18} (Fig. 13.3) was created ([4, 12], 2013). The main feature of an author method of pyrolysis (AMP) is that the composition of the resultant products depends on not only the reactionary conditions (temperature, reagent concentration and time of them stay in the most high-temperature zone of

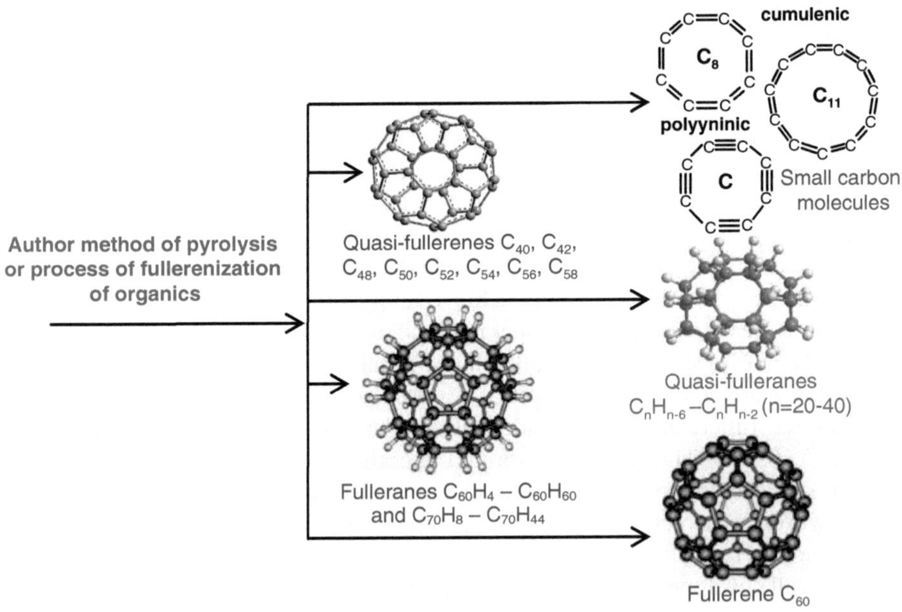

Fig. 13.3 Condensed products obtained by author's method pyrolysis of organics (benzene, toluene, xylene, pyridine and ethanol)

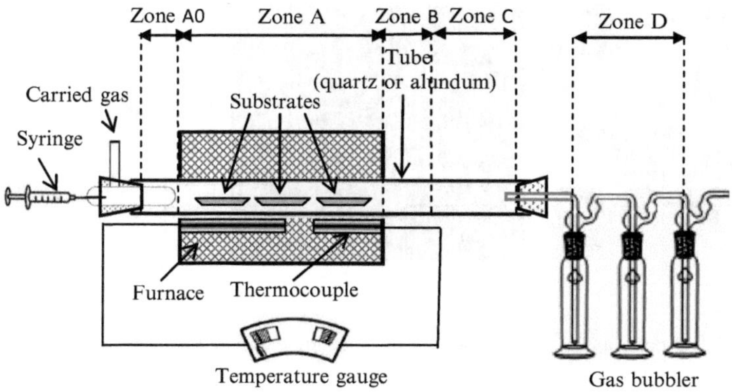

Fig. 13.4 A scheme of the reactor for new (author's) method of pyrolysis as process of fullereniization of molecules of organic compounds

the reactor), but also especially the place of localization of these products in the reaction space (Fig. 13.4). At AMP the resultant condensed substances and a part of soot in a gas reactionary flow are taken out from the high temperature (950–1100 °C) zone **A** (Fig. 13.4) into a low temperature (<300 °C) zones B and C of reactionary space. The most part of pyrolytic soot containing various carbon nanostructures (fibres, nanofilaments, nanotubes and onions [12]), is deposited (as well as in typical CVD process) in a high-temperature zone **A**. In low temperature zones, as for the first time it is revealed by us [4, 12], the carbon molecules of all types are condensed from a reactionary flow, which the nucleation and growth is fulfilled in zone **A** and proceeds in a gas phase of reactionary space between zones **A**, **B** and **C** (Fig. 13.4). Therefore it is possible to consider that the carbon molecules and them hydrides are formed at the expense of radical reactions, in particular, polymerization and polycondensation of molecules of precursor and fragments of them destruction.

For preparation of samples for study 8–10 tests under similar reactionary conditions usually realized. The products located in each of low temperature zones **B**, **C** and **D** (Fig. 13.4) mixed and then subjected to careful processing by benzene and xylene for more complete extraction from soot of the condensed substances. Benzene (or xylene) extracts concentrated and by means of method of salting out by an ethanol received powdery deposits, which carefully dried and then investigated by different methods, such as temperature-programmed desorption mass spectrometry (TPDMS), NMR and IR-spectroscopy, chemical and X-ray analysis. According to the chemical analysis all products, deposited by spirit, alongside with carbon and oxygen always contain hydrogen, but contain it in different quantities. The contents of hydrogen in products of pyridine fullerenization can be from 3.9 up to 5.8 mass%, whereas in products of benzene fullerenization – from 2.8 up to 4.9 mass%. All products of fullerenization contain also oxygen up to 3.1 mass%. Deposited amorphous samples (Fig. 13.5) with the large contents of hydrogen have slightly yellowish, and with smaller – reddish colour (Fig. 13.5).

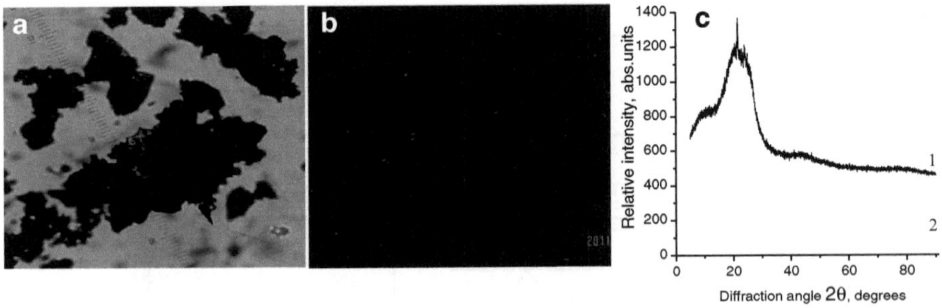

Fig. 13.5 XRD patterns (**c**), of and optical microscopy deposited samples slightly *yellowish* (**a**) and *reddish* (**b**) colour

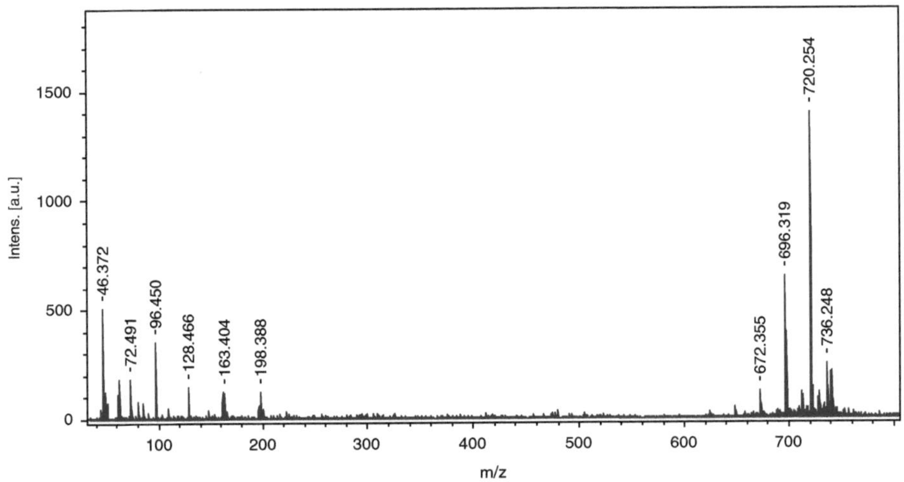

Fig. 13.6 Mass spectrum of condensed product **C** of fullerenization of benzene

Notably, that the composition of the condensed substances in zones **B**, **C** and **D** is appreciable different. However products of AMP of different substances always there are carbon molecules and their hydrides.

The study of the condensed products of fullerenization of molecules of benzene, toluene, xylene, pyridine and ethanol by a method of MALDI (matrix-assisted laser (nitrogen, 337 nm) desorption/ionization) has shown [4, 12], that at their mass spectra there are peaks with meanings of m/z, the majority from which were not detected at all earlier (except for, certainly, carbon plasma). So products of fullerenization of benzene alongside with fullerene C_{60} (Fig. 13.6) contain fulleranes of different composition (Fig. 13.7) and, that especially is important, completely saturated by hydrogen fullerane $C_{60}H_{60}$. It is important to note, that earlier fullerane of equiatomic composition ($C_{60}H_{60}$) was not detected at all in one of products of heterophase hydrogenation of fullerite or fullerene. Moreover, in

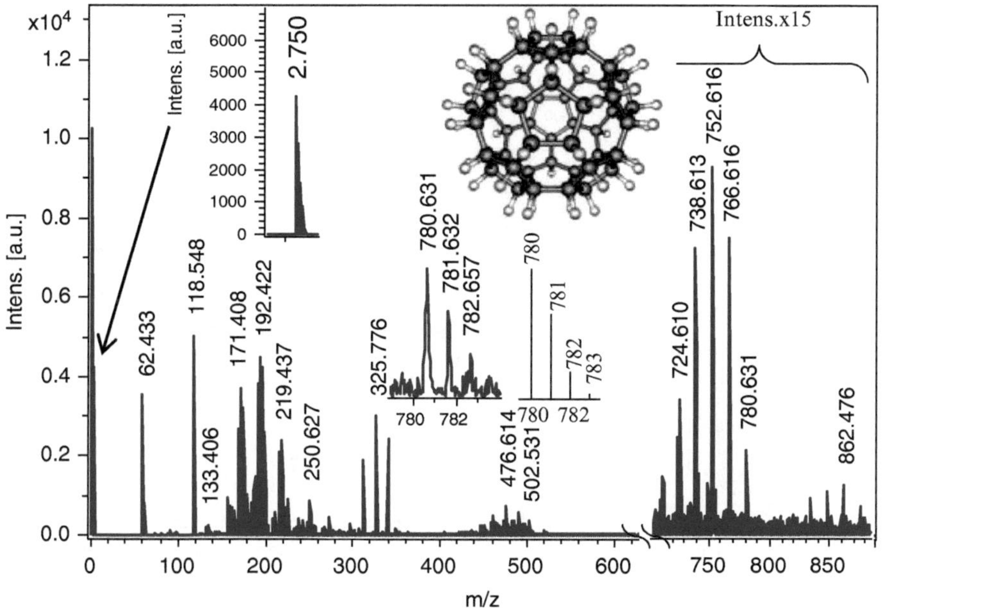

Fig. 13.7 Anions mass spectrum of condensed product **B** of fullerenization of benzene with the expansion around the m/z 2–4; m/z 780 peaks and the calculated isotope mass ratios for molecule of fullerane ($C_{60}H_{60}$) in the insets

products of fullerenization the rather intensive peaks are detected with meaning of m/z 480, 504 and 576 (Fig. 13.8), which thin structures will be satisfactorily coordinated to counted isotope distribution for molecules of quasi-fullerenes C_{40}, C_{42} and C_{48}. Notably, that just in products of fullerenization, for example, toluene (Fig. 13.9a) simultaneously with carbon molecules the rather large number of peaks (Fig. 13.9a, b) is detected with meanings of m/z from 254 up to 574, which strictly differ on 24 or 26 units.

According to thin structure these peaks correspond to different composition of quasi-fulleranes from $C_{20}H_{14}$ up to $C_{46}H_{22}$. Especially it is necessary to note, that at fullerenization of a heteroatomic molecule of pyridine (C_5H_5N) are formed not only heteroatomic molecules as exohedral hydrogenated and hydroxylated azafullerenes $(C_{35}N_5)H_9$, $(C_{45}N_5)(OH)_3H_{14}$ and $(C_{49}N_{11})(OH)_5H_{18}$ (Fig. 13.10) but also homoatomic molecules C_{60} and C_{48} as well as C_3-C_{18} (Fig. 13.8) [4, 12]. The fact of formation of carbon molecules from pyridine molecules confirms importance of a stage of dehydrogenated destruction of precursor molecules at growth of fullerene and quasi-fullerene.

In powdery products of fullerenization, for example pyridine, are detected also even and odd ions of small carbon molecules C_3-C_{20} (Fig. 13.8), which were earlier found out in mass spectra of carbon plasma. Herein it is important to note that small carbon molecules such as C_2 and C_3, C_4 and C_5 together with polyynes (HC_nN,

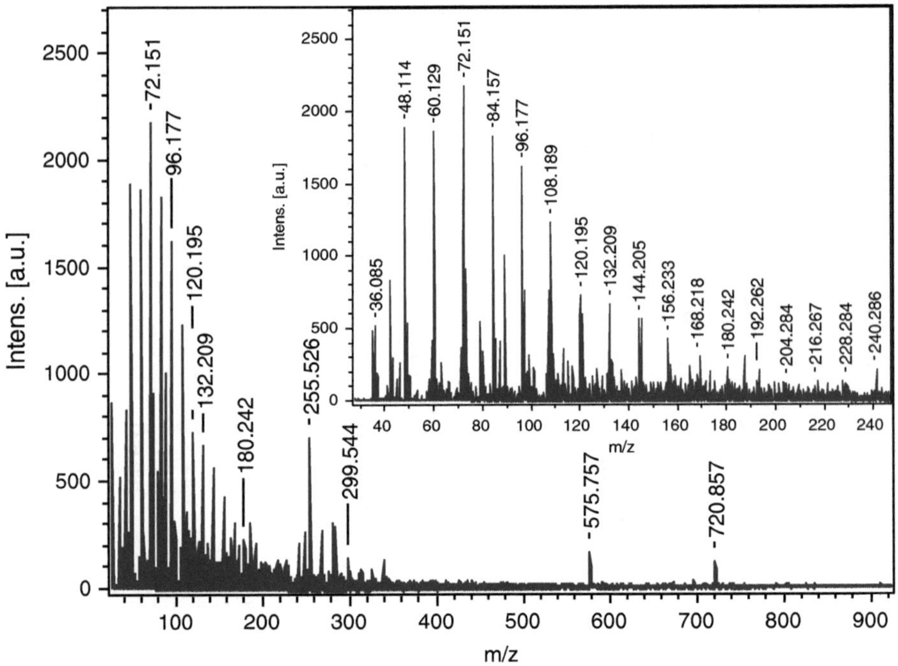

Fig. 13.8 Anions mass spectrum of product **B** of pyridine fullerinization with the expansion around the m/z 25–250 region

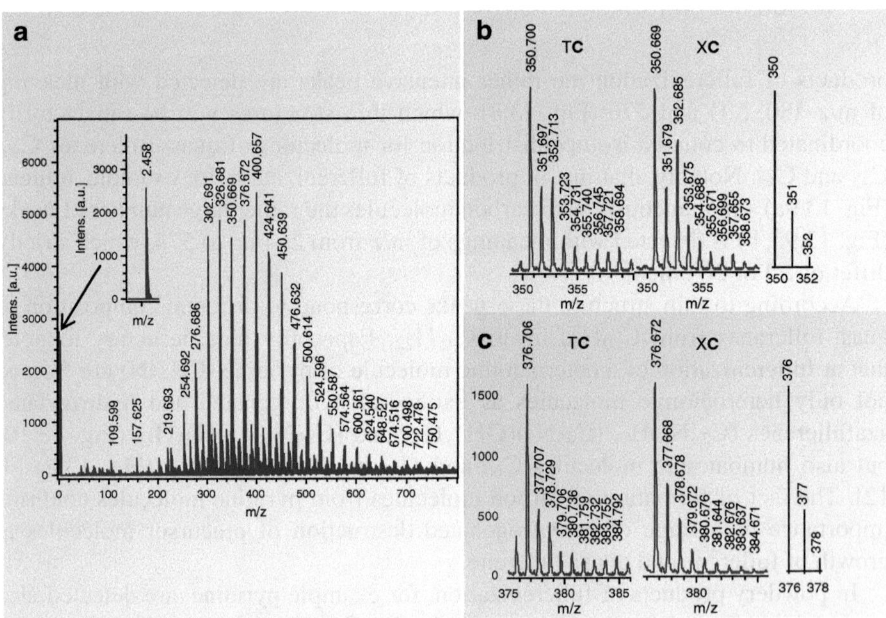

Fig. 13.9 Cations mass spectra of the products **C** of fullerenization of toluene (**a**); expansions around m/z 350 (**b**) and 376 (**c**) peaks in the cations mass spectra of the product **C** of fullerenization of toluene (*left*) and xylene (*centre*) and the calculated isotope mass ratio for $C_{28}H_{14}$ and $C_{30}H_{16}$ molecules respectively

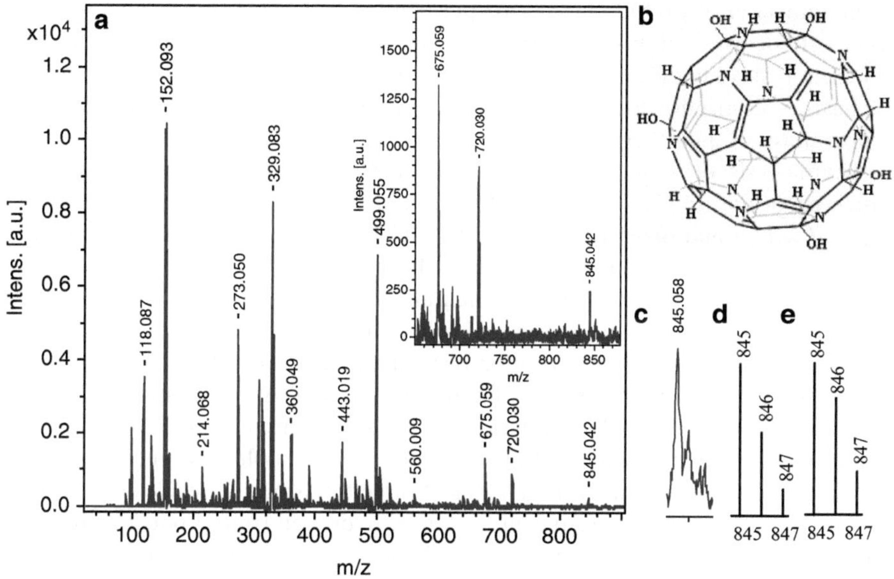

Fig. 13.10 Mass spectrum of product **C** of pyridine fullerenization containing heteroatomic fullerene-like molecules of exohedral polyazafullerenes $(C_{35}N_5)H_9$, $(C_{45}N_5)(OH)_3H_{14}$ and $(C_{49}N_{11})(OH)_5H_{18}$ (**a**), scheme of molecule $(C_{49}N_{11})(OH)_5H_{18}$ (**b**), the expanded patterns of the 845 m/z peak (**c**), the calculated isotope mass ratios for $(C_{49}N_{11})(OH)_5H_{18}$ (**d**) and $C_{70}H_5$ (**e**) molecules

were $n \leq 11$) only in the circumstellar medium [25] are found out. In laboratory conditions these carbon molecules only in solid argon about 15K were isolated. However a time of life of such frozen clusters is extremely small (~10 ms).

Detected in products of fullerenization of benzene, toluene and pyridine of the small carbon molecules (Fig. 13.9) give the basis to assume, that to destruction of molecules, in particular, C_5H_5N, precedes their complete dehydrogenation with formation of ring molecules C_6 (polyynic or cumulenic structure), which also easily break up on clusters of the smaller sizes (C_3-C_5). Therefore clusters C_3–C_5 are detected in products of fullerenization of molecules both benzene [4, 13, 26–30], and pyridine [12, 26, 28, 31–35]. Fullerene C_{60} and quasi-fullerenes, probably, are formed at the expense of reactions of polycondensation and polymerization both molecules of precursor, and products it destruction.

One of the basic obstacles for successful use in hydrogen power of system fullerene-fullerane-hydrogen-fullerene is that at temperatures of fulleranes dehydrogenation is observed also the destruction of them fullerene cage. However at the possibility of synthesis of quasi-fulleranes (or fulleranes) from benzene (or ethanol) the situation can cardinal change, as such system as ethanol – quasi-fullerane (fullerane) – hydrogen can be more effective. It is known [8] that the intensive loss of mass at destruction of fullerane $C_{60}H_{36}$ in the process of its dehydrogenation essentially depends on a method of their preparation. Therefore the study of process of thermostability of fulleranes and quasi-fulleranes is extremely important for successful decision of this problem. The composition of gaseous phase of products of thermolysis of hydrogenated carbon molecules is expedient for analyzing only on the basis of desorption mass spectra which have been written down in all a range of temperatures of their thermostability.

Thermolysis of synthesized by means of AMP of the condensed products, which according to the data NMR, IR and mass spectrometry contain fulleranes $C_{60}H_4$-$C_{60}H_{60}$ and quasi-fulleranes C_nH_{n-6}-C_nH_{n-2} ($n = 20$–54), was investigated in an interval of temperatures 30–750 °C. From thermodesorption curves can see that all of them contain hydrogen, which allocation is realized in all investigated temperature interval. From products of fullerenization, in particular, benzene and pyridine the rather intensive allocation of hydrogen begins with 25 °C (Fig. 13.11, inset). However under the contents of hydrogen products of fullerenization benzene (Fig. 13.11, curves 3 and 4) and pyridine (Fig. 13.11, curves 1 and 2) appreciably differ by comparison of identical (3 mg) qualities of samples: product of pyridine fullerenization contains of hydrogen almost three times more. Besides temperature of a beginning of more intensive allocation of hydrogen from product of pyridine fullerenization also is a little bit lower. It is possible to assume that heteroatomic fullerene-like molecules (polyazafullerenes [12, 31]), which are detected in product of pyridine fullerenization is less thermostable.

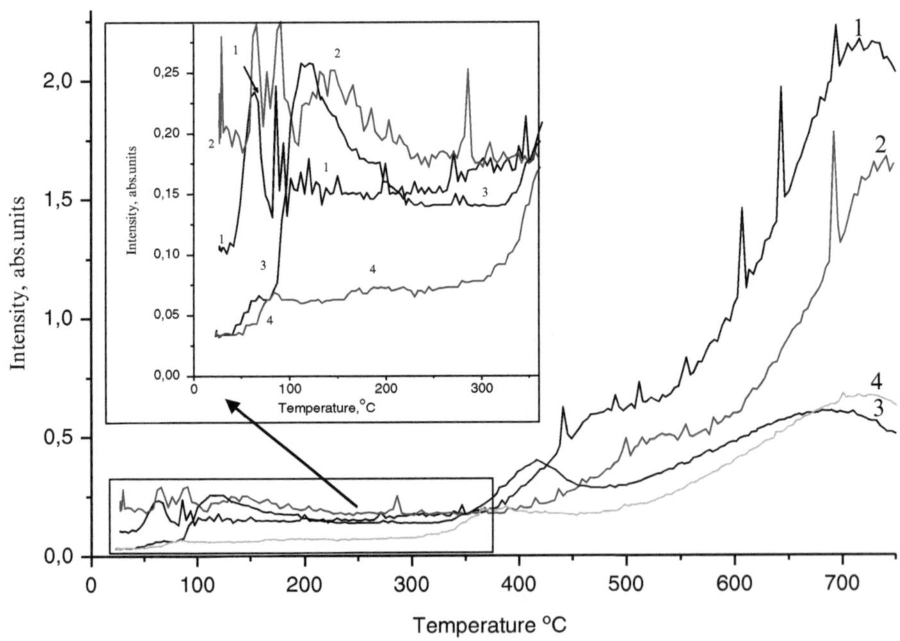

Fig. 13.11 Experimental thermodesorption *curves* for the 3 amu peaks for samples of quasi-fulleranes (1 and 2) and fulleranes (3 and 4) obtained from pyridine and benzene respectively with 25–350 °C region in the inset

Hence, as against fulleranes synthesized at fullerenes hydrogenation fulleranes and quasi-fulleranes as the products of homogeneous processes begin to dehydrogenate at lower temperatures. Probably, it is connected with greater dispersity of particles of products of fullerenization or greater propensity of quasi-fulleranes molecules to dehydrogenation to formation more stable at the raised temperatures of molecules of quasi-fullerenes.

13.4 Conclusions

New (authors) method of pyrolysis as the process of fullerenization (thermal (~1000 °C) transformation molecules of organic substances into carbon molecules and their hydrides) is created. First, in contrast to already known methods for producing of fullerenes and fulleranes the reactionary conditions of a fullereniization of molecules of benzene, toluene and pyridine into the molecules not only fullerene C60, fulleranes, but also quasi-fullerenes C_{40}, C_{42}, C_{48}, quasi-fulleranes C_nH_{n-6}-C_nH_{n-2} (n = 20–46) and small carbon molecules C_3–C_{18} are developed.

Only in powdery products of fullerenization by a mass spectrometric method the cluster with m/z 780 was detected, which the thin structure of spectrum completely corresponds to fullerane of equiatomic composition, $C_{60}H_{60}$. First fulleranes $C_{60}H_8$–$C_{60}H_{60}$ and $C_{70}H_8$–$C_{70}H_{44}$ were synthesized not from fullerene (or fullerite).

First in mass spectrum of powdery product of pyridine fullerenization are detected new heteroatomic fullerene-like molecules, such as exohedral polyazafullerenes polyazafullerenes $(C_{35}N_5)H_9$, $(C_{45}N_5)(OH)_3H_{14}$ and $(C_{49}N_{11})(OH)_5H_{18}$.

References

1. Haufler RE, Conceicao J, Chibante LPF et al (1990) Efficient production of C_{60} (buckminsterfullerene), $C_{60}H_{36}$, and the solvated buckide ion. J Phys Chem 94(24):8634–8636
2. Cataldo F, Iglesias-Groth S (2010) Fulleranes: the hydrogenated fullerenes. Springer, Dordrecht, p 278
3. Goldshleger NF, Moravsky AP (1997) Hydrides of the fullerenes. Usp Khim 66(4):353–375
4. Kharlamov AI, Bondarenko ME, Kirillova NV (2012) New method for synthesis of fullerenes and fullerene hydrides from benzene. Russ J Appl Chem 85(2):233–238
5. Luzan SM, Tsybin YO, Talyzin AV (2011) Reaction of C_{60} with hydrogen gas: in situ monitoring and pathways. J Phys Chem C 115(23):11484–11492
6. Luzan S (2012) Materials for hydrogen storage and synthesis of new materials by hydrogenation. PhD thesis, Umeå University, Umeå, p 85
7. Peera A, Saini RK, Alemany LB et al (2003) Formation, isolation, and spectroscopic properties of some isomers of $C_{60}H_{38}$, $C_{60}H_{40}$, $C_{60}H_{42}$ and $C_{60}H_{44}$. Eur J Org Chem 21:4140–4145
8. Peera AA (2004) Fullerene hydrides and studies toward the synthesis of fulvalenes. PhD thesis, Houston

9. Shigematsu K, Abe K, Mitani M, Tanaka K (1993) Catalytic hydrogenation of fullerenes in the presence of metal catalysts in toluene solution. Fullerene Sci Technol 1(3):309–318
10. Talyzin AV, Dzwilewski A, Sundqvist B et al (2006) Hydrogenation of C_{60} at 2 GPa pressure and high temperature. Chem Phys 325(2):445–451
11. Zhang JP, Wang NX, Yang YX, Yu AG (2004) Hydrogenation of [60] fullerene with lithium in aliphatic amine. Carbon 42(3):667–691
12. Kharlamova G, Kharlamov O, Bondarenko M (2013) Hetero-carbon: heteroatomic molecules and nano-structures of carbon. Chap. 31. In: Vaseashta A, Khudaverdyan S (eds) Advanced sensors for safety and security, NATO science for peace and security series B: physics and biophysics. Springer, Dordrecht, pp 339–357
13. Kharlamova G, Kharlamov O, Bondarenko M (2015) Nanosensors in systems of ecological security. In: Bonča J, Kruchinin S (eds) Nanotechnology in the security systems. NATO science for peace and security series C: environmental security. Springer, Dordrecht, pp 231–242, Chap. 20
14. Kroto HW, Heath JR, O'Brien SC et al (1985) C_{60}: buckminsterfullerene. Nature 318:162–163
15. Rohlfing C, Kaldor J (1984) Production and characterization of supersonic carbon cluster beams. Chem Phys 81:3322–3330
16. Siegmann K, Hepp H, Sattler K (1995) Multiphoton ionization mass spectroscopy of fullerenes in methane diffusion flames. Mat Res Soc Proc 359:517
17. Siegmann K, Hepp H, Sattler K (1996) High-resolution height-profile analysis and laser-ionization characterization of a wide range of fullerenes in laminar diffusion flames. Surf Rev Lett 3:741
18. Wurz P, Lykke KR, Pellin MJ, Gruen DM (1992) Characterization of fullerenes by laser-based mass spectrometry. Vacuum 43(5–71):381–385
19. Kratschmer W, Lamb LD, Fostiropoulos K, Huffman DR (1990) Solid C_{60}: a new form of carbon. Nature 347:354–358
20. Howard JB, McKinnon JT, Makarovsky Y et al (1991) Fullerenes C_{60} and C_{70} in flames. Nature 352:139–141
21. Kroto HW (1990) C_{60} fullerenes, giant fullerenes and soot. Pure Appl Chem 62:407–415
22. Paquette LA, Ternansky RJ, Balogh DW, Kentgen G (1983) Total synthesis of dodecahedrane. J Am Chem Soc 105(16):5446–5450
23. Henderson CC, Cahill PA (1993) $C_{60}H_2$–synthesis of the simplest C60 hydrocarbon derivative. Science 259(5103):1885–1887
24. Ballenweg S, Gleiter R, Kratschmer W (1993) Hydrogenation of buckminsterfullerene C-60 via hydrozirconation – a new way to organofullerenes. Tetrahedron Lett 34(23):3737–3740
25. Cataldo F (2004) Cyanopolyynes: carbon chains formation in a carbon arc mimicking the formation of carbon chains in the circumstellar medium. Int J Astrobiol 3:237–246
26. Kharlamov A, Bondarenko M, Kharlamova G (2013) Mass spectrometric research of hydrogenated molecules of carbon as products of pyrolysis of benzene and pyridine vapours. Chem Mater Eng 1(4):122–131
27. Kharlamov A, Bondarenko M, Kharlamova G (2013) Hydrogenated molecules of carbon as products of new pyrolysis method of toluene, xylene and ethanol. Univers J Chem 1(3):102–112
28. Kharlamov A, Bondarenko M, Kharlamova G et al (2013) A new method of synthesis carbon with onion-like structure with high (10–13 %) content of nitrogen from pyridine. Univers J Mater Sci 1(2):78–86
29. Kharlamov A, Kharlamova G, Bondarenko M, Fomenko V (2013) Joint synthesis of small carbon molecules (C_3–C_{11}), quasi-fullerenes (C_{40}, C_{48}, C_{52}) and their hydrides. Chem Eng Sci 1(3):32–40
30. Kharlamov A, Kharlamova G, Bondarenko M, Fomenko V (2013) New method for generation of carbon molecules and clusters. Open J Synth Theory Appl 2(1):38–45
31. Kharlamov AI, Kharlamova GA, Bondarenko ME (2013) New products of a new method for pyrolysis of pyridine. Russ J Appl Chem 86(2):167–175

32. Kharlamov AI, Kharlamova GA, Bondarenko ME (2013) New low-temperature method for joint synthesis of C_{60} fullerene and new carbon molecules in the form of C_3–C_{15} and quasi-fullerenes C_{48}, C_{42}, C_{40}. Russ J Appl Chem 86(8):1174–1183
33. Kharlamov AI, Kharlamova GA, Bondarenko ME (2013) Preparation of onion-like carbon with high nitrogen content (~15 %) from pyridine. Russ J Appl Chem 86(10):1493–1503
34. Kharlamov O, Kharlamova G, Bondarenko M, Fomenko V (2013) Small carbon molecules and quasi-fullerenes as products of new method of hydrocarbons pyrolysis. Chap. 30. In: Vaseashta A, Khudaverdyan S (eds) Advanced sensors for safety and security, NATO science for peace and security series B: physics and biophysics. Springer, Dordrecht, pp 329–338
35. Kharlamov O, Bondarenko M, Kharlamova G et al (2015) Nanoecological security of foodstuffs and human. Chap. 19. In: Bonča J, Kruchinin S (eds) Nanotechnology in the security systems, NATO science for peace and security series C: environmental security. Springer, Dordrecht, pp 215–229

Chapter 14
Hyperspectral Stochastic Optical Reconstruction Raman Microscopy for Label-Free Super-Resolution Imaging Using Surface Enhanced Raman Spectroscopy

Aykutlu Dâna

Abstract Super-resolution imaging is an emerging field that has attracted attention in the recent years due to far the reaching impact in biology. All super-resolution techniques use fluorescent labels to image nanoscale biomolecular structures. In contrast, label-free nanoscopic imaging of the chemical environment of biological specimens would readily bridge the supramolecular and the cellular scales, if a chemical fingerprint technique such as Raman scattering can be coupled with super-resolution imaging, overcoming the diffraction limit. In order to achieve this goal, we propose to develop a super-resolved stochastic hyperspectral Raman microscopy technique for imaging of biological architectures. The surface enhanced Raman spectroscopy (SERS) signal contains information about the presence of various Raman bands, allowing for the discrimination of families of biomolecules such as lipids, proteins, DNA. The rich, fluctuating spectral information contained in the single molecule SERS signal possesses a great potential in label-free imaging, using stochastic optical reconstruction microscopy (STORM) methods. In a recently published work, we demonstrated 20 nm spatial resolution using the spectrally integrated Raman signal on highly uniform SERS substrates. A mature version of our method would require development of spectrally resolved nanoscale Raman imaging. Development of stochastic Raman imaging addresses the issue by design and construction of a Raman microscope with hyperspectral imaging capability that will allow imaging of different Raman bands of the SERS signal. Novel computational techniques must also be developed that will enable extraction of hyperspectral STORM images corresponding to different Raman bands, while simultaneously allowing conventional STORM data to be collected using the well-established labelling techniques. The resulting technique (Hyperspectral Raman

A. Dâna (✉)
UNAM Institute of Materials Science and Nanotechnology, Bilkent University, 06800 Ankara, Turkey
e-mail: aykutlu@unam.bilkent.edu.tr

T.A. Camesano (ed.), *Nanotechnology to Aid Chemical and Biological Defense*, NATO Science for Peace and Security Series A: Chemistry and Biology,
DOI 10.1007/978-94-017-7218-1_14

STORM or HyperSTORRM) has the potential to complement the available labeled stochastic imaging methods and enable chemically resolved nanoscopy.

14.1 Introduction

In this chapter, we discuss issues relating to the development of an advanced imaging technique that would allow nanometer scale resolution far-field imaging of biological architectures with the chemical specificity present in the Raman scattered optical signal. Specifically, the method uses stochastic reconstruction methods on the fluctuating optical signal present in single-molecule surface-enhanced Raman phenomena to achieve this goal.

Raman microscopy can be applied in a confocal configuration, to provide Raman maps of surfaces. Raman microscopy using pulsed illumination resulted in CARS (coherent anti-stokes Raman) and SRS (stimulated Raman) microscopy, also typically exhibiting diffraction limited resolution [1]. Similarly, confocal or wide-field Raman imaging displays a resolution of $\sim$250 nm at best. Near-field microscopy using tip-enhanced Raman microscopy provides nanometer scale resolution, however, the technique is slow and is not far-field, because it requires nanometer separation between a scanning tip and the sample, and can't be applied for 3D imaging [2]. Plasmonic structures made the acquisition of Raman spectra possible from sub-monolayer coverage molecular films, a technique known as Surface-Enhanced Raman Spectroscopy (SERS) [3–5]. Using SERS, single molecule level sensitivity can be achieved [6–9]. This was previously attributed to extremely high enhancement factors of 10^9–10^{12}. It was later claimed that much lower enhancement factors (10^6–10^8) may be sufficient for such observations [10]. Stochastic optical reconstruction microscopy (STORM) is a versatile technique that allows sub-diffraction limited resolution through mathematical analysis of a video sequence of blinking fluorescence signals in a sample. The centroids of blinking spots are calculated using various algorithms to achieve the reconstruction. Traditionally, photoactivated or self-blinking fluorescent molecular labels are used in STORM to highlight various cellular architectures. Labeling allows discrimination of components, however potentially disturbs the native state of the biomolecular structures. The SERS signal, at single-molecule level, typically shows a rapid blinking behavior which can be captured with a camera or spectrometer. This non-stationary nature of the SERS signal and the rich spectral information contained in the acquired spectra, in principle, provides an opportunity to exploit the SERS signal for chemically-resolved, label-free, super-resolution stochastic imaging [11]. Wide-field imaging and stochastic reconstruction methods have been applied to observe the spatial distribution of the of the SERS signal, allowing reconstruction of the field distributions of individual hot-spots [12]. Traditionally, similar reconstruction methods have used fluorescence signal from single molecules and nanoparticles, relying on photo-activation, photo-bleaching, and molecular diffusion [13]. An increasing number of articles used wide-field imaging and stochastic analysis of the SERS signal as an application of SERS in super-resolved imaging of field distributions hot-spots [14]. However, for SERS to be a viable way of reconstructing the nanoscale structure of a biomolecular layer, rather than the field distribution of the underlying hot-spot itself, a large number of closely packed hot-spots is required within a focal spot with an area defined by the diffraction limit. In this work, we take a completely novel approach to Raman imaging. Previously, meta-surfaces with periodicities larger than the diffraction limited spot size have been shown to produce high and uniform enhancement [15]. Recently, we used an engineered meta-surface with sub-wavelength periodicity to demonstrate high uniformity and high spatial resolution confocal SERS imaging [16]. In following work, the plasmonic designs are extended and techniques for production of easy-to-fabricate SERS substrates are developed that exhibit a high density of hot-spots while allowing single-molecule level Raman spectroscopy [17, 18]. SERS and STORM techniques are combined to produce reconstructed images of biomolecular structures. The enhancement uniformity and performance featured by the substrate is sufficient to perform STORM with 20 nm resolution [17]. In this chapter, hyperspectral stochastic imaging using the SERS signal and the development of a label free nanoscopic imaging technique is discussed.

14.2 Basic Description of the Hyperspectral Super-Resolved Raman Imaging System

Figure 14.1 shows the main idea of a Hyperspectral Stochastic Optical Reconstruction Raman Microscopy (HYPERSTORRM). A hyperspectral wide-field Raman imaging system is described that can record SERS blink events with distinct spectral characteristics, belonging to different vibrational bands of biomolecules. Advanced stochastic reconstruction algorithms are needed to obtain nanometer resolution images that show chemical composition of biological architectures with nanometer resolution. In this imaging mode, the samples are placed on highly-uniform SERS substrates that exhibit single-molecule level sensitivity. In the basic version of the HYPERSTORRM microscope, a fixed excitation wavelength is used, along with multiple cameras (multispectral imaging module). Four Raman bands corresponding to common Raman active modes of biological structures was be selected during an experiment (by proper choice of dichroic mirrors and narrowband filter wavelengths) and initial images contained chemical information based on four Raman frequencies. In the second variation of the microscope, a hyperspectral imaging module is used (optical scheme shown in Fig. 14.2). The advantage of using a hyperspectral module is recording of the complete SERS spectrum, enabling stochastic reconstructed imaging using all available Raman information. Pulsed (pico-nanosecond pulse width) source can be used to reduce sample damage.

In hyperspectral image acquisition using a 2D detector array, typically a 1D grating with a linear slit, is used to separate the information into one dimensional

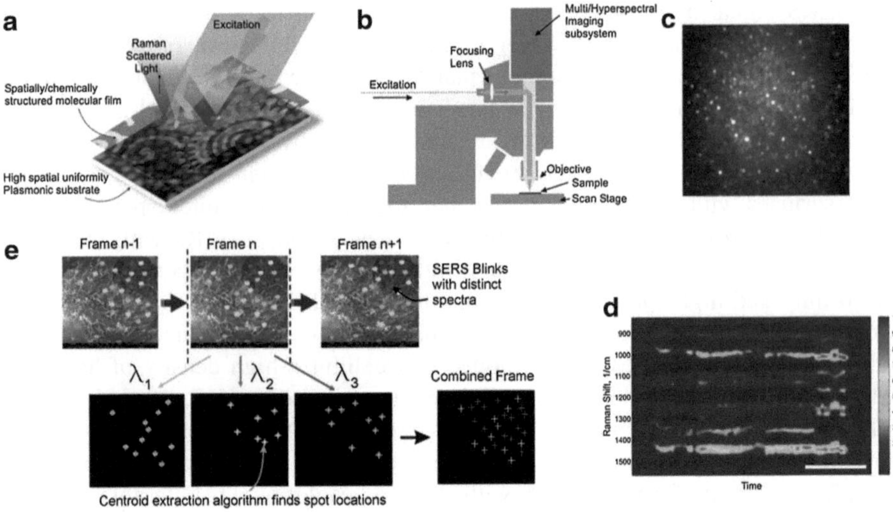

Fig. 14.1 Description of the HYPERSTORRM Microscope. (**a**) A biological sample (such as a cell membrane, self-assembled supramolecular structures, extra or intracellular matrix components) is placed on a high performance and high uniformity plasmonic substrate. Under wide-field excitation, SERS signal originates from the sample. (**b**) A Raman microscope system equipped with a multi/hyperspectral imaging module allows collection of light from the sample. (**c**) Example wide-field SERS image from a uniform plasmonic substrate covered with a uniform peptide layer is given, showing evenly distributed *blinking spots*. (Supporting information of [17] has videos showing SERS induced *blinking* in various samples). (**d**) Spectrum collected from a single *blinking spot* using a fiber shows typical characteristic fluctuations in the intensity and frequency of Raman bands during single-molecule *blinking* events. (**e**) HYPERSTORRM uses such fluctuations and their correlations to reconstruct a Raman map with nanometer scale resolution

spatial and spectral directions [19]. As the sample is scanned in one spatial direction, a hyperspectral map can be acquired. This conventional method of hyperspectral imaging is not adequate for stochastic super-resolution imaging. Instead, a way of reducing the three dimensional (two spatial and one spectral) data to the two dimensional space of the detector must be developed, without resorting to sample scanning. One has to either use a large number of spectrally separated cameras, or approximate the solution by imposing limitations on the optical signal. Indeed, it is shown in Sect. 4 that, by limiting the spatial extent of the image and spectral extent of the signal and by introducing the correlations present in fluctuations of the SERS signal, the problem of hyperstactral reconstruction can be solved. In order to clarify our approach, consider the scheme presented in Fig. 14.2, where the image is recorded using a 2D transmission grating in front of the image sensor, and 0th and higher order images are acquired. The center (0th order) image is used for localization. Using the localization data and by using time correlations of spectral features with the 0th order image pixels, one can extract the spectral content of the SERS signals and assign them to spots in the 0th order image. In this way, the hyperspectral super-resolution maps can be produced from SERS data.

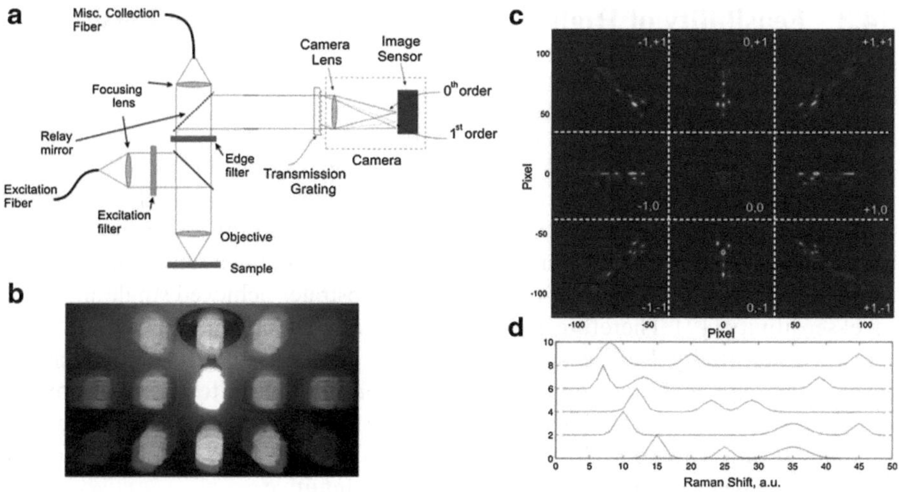

Fig. 14.2 Preliminary optical design for HYPERSTORRM imaging. (**a**) A Raman microscope is modified to incorporate a high sensitivity camera and high efficiency transmission grating to form a hyperspectral imaging system. (**b**) The 0th order of the grating produces a real image and higher orders produce spectrally resolved images. Example image of a fluorescent light bulb taken with a 2D grating in front of a camera is shown. (**c**) Simulation of imaging of a spectrally rich sample is shown. Five spots in the form of a 'T' (*center of image*) are assumed to have different spectra. The 2D grating separates the spectra in positive and negative orders. By limiting the field of view to a diameter of about 10 μm and spectra to visible range, 0th and 1st order images are well separated. The 0th order image can be used for localization and spatial reconstruction. The 1st order images contain spectral information, which can be used by advanced mathematical methods for the extraction of spectral data. Mathematically it is not possible to fully invert the imaging scheme to reconstruct the spectral content. However, by introducing time as a variable, the problem can be inverted to give the spectra corresponding to each spot. (**d**) The hypothetical spectra of the scattering centers used in the simulation in (**c**) are shown. Such spectral separation necessitates the use of a high sensitivity camera, as the optical signal is spatially multiplexed in this scheme

The design can be built around a commercial microscope system, allowing confocal Raman microscopy. The microscope may be modified for excitation with multiple wavelengths, including continuous and pulsed laser sources. The multi-spectral imaging head can be designed to allow easy change of filter combinations for use with different excitation wavelengths. A careful initial study must be conducted for determination of common Raman bands that are present in samples of interest (cellular components, proteins, peptidic nanostructures, DNA, lipids etc.). The laser excitation sources must be chosen to minimize fluorescence effects, if possible. Once the Raman bands and excitation wavelengths are determined (important for first generation experiments), filter wavelengths are fixed. Initial phase of design must focus on optimization of the detection efficiency, while matching the microscope and Electron Multiplied (EM) CCD cameras, in terms of aperture and pixel size. Synchronization of the frames from multiple cameras at high frame and data capture rate needs to be achieved by using high performance multicore desktop computers.

14.3 Feasibility of Highly-Uniform Single-Molecule Level SERS Substrates

Previous work on the design and fabrication of high-uniformity and single-molecule level sensitivity SERS substrates, necessary for the realization of HYPERSTORRM, demonstrates the feasibility of the approach presented here. In initial experiments with quarter micron scale plasmonic structures, high-uniformity SERS spectroscopy and confocal imaging were demonstrated in a recent publication [16]. A further scaled down version of high-uniformity SERS substrates achieved single-molecule level sensitivity [17]. Therefore, the feasibility of fabricating high spatial uniformity and high sensitivity Raman substrates is well demonstrated. It is necessary to investigate ways to further improve hot-spot density and uniformity. An obvious direction is further scaling down the plasmonic surfaces to give a hot spot spacing of about 10 nm. However, this scaling would require advances in fabrication, and keeping the enhancement factors above 10^8 is not intuitive.

The improvement of spatial uniformity and smaller characteristic dimension SERS substrates, while increasing the field enhancement factor, plays a crucial role in achieving HyperSTORRM images with improved signal-to-noise ratio and resolution. Previously, Metal-Insulator-Metal (MIM) structures that exhibit spatially uniform SERS enhancement were studied. These were fabricated by using electron beam lithography as shown in Fig. 14.3. A detailed study of the relation between geometry and plasmonic enhancement properties has been conducted, using computational tools as well as analytical calculations based on circuit models of such structures (Fig. 14.3c). The models are observed to reproduce experimental results fairly well (Fig. 14.3d), and a high spatial uniformity can be achieved on large scales (>250 nm, Fig. 14.3e, f). Nanoscale uniformity is harder to achieve in the given geometry, and there is inherent non-uniformity within a unit-cell of the surface (Fig. 14.3g). However, as the characteristic size of the plasmonic structures shrink, it is experimentally observed that a more uniform enhancement is present. The uniformity issue is not straight forward and deserves more careful study.

In our model plasmonic system, the primary requirement for high SERS enhancement is the presence of plasmonic resonances at both the excitation and scattering wavelengths. Also, for high resolution imaging, the structures are required to be omnidirectional, allowing use of high numerical aperture objectives. A simplified analytical model of the periodic MIM structure allows rapid estimation of the effect of geometrical parameters on resonant properties (Fig. 14.3c). It can be seen that, scaling the MIM widths down to 10 nm while keeping the resonances at given laser wavelengths (532–785 nm) would require thin dielectric layers to the order of few nanometers. Our process uses atomic layer deposition to achieve high uniformity and excellent thickness control, therefore we have been able to shrink down to about 30 nm island size (inset of Fig. 14.3g). The improved surfaces resulted in the collection of preliminary images shown in Fig. 14.4.

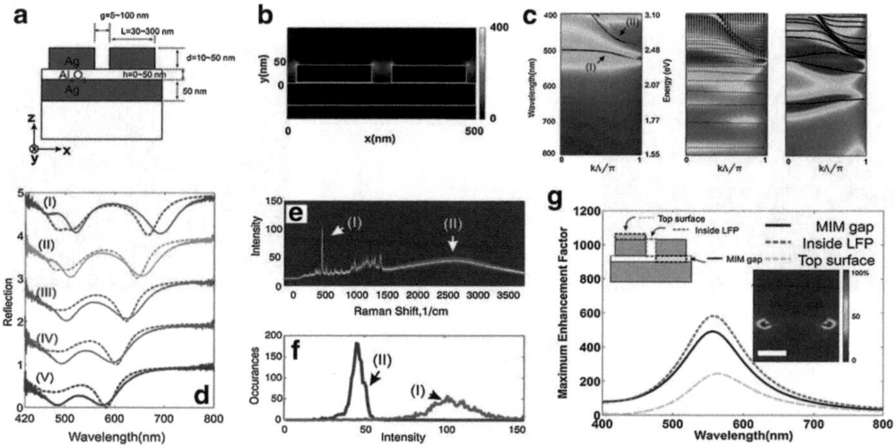

Fig. 14.3 Engineering high-performance plasmonic substrates. (**a**) Basic Metal-Insulator-Metal (*MIM*) structure used for high uniformity SERS enhancement. (**b**) Typical enhancement profile (shown for a structure with diminishing dielectric thickness). (**c**) Band diagrams calculated for 0, 3 nm and 10 nm dielectric thickness for 250 nm period structures. (**d**) Calculations and measurements of reflectance on surfaces prepared by electron beam lithography show good agreement. (**e**) 1600 SERS spectra collected over 10 × 10 μm area show excellent repeatability. (**f**) Histograms show 10–20 % uniformity of enhancement. (**g**) Shrinking the dimensions further down improves field enhancement and uniformity (inset shows fields for 40 nm period structures, scale bar 10 nm). Enhancement factor shows a non-uniform distribution within the unit cell of the plasmonic surface, which prompts investigation of alternative structures or geometries

The surfaces are fabricated by spontaneous self-organization of Ag nanosilands during deposition of ultrathin Ag films on oxides. Due to the de-wetting of Ag, the films do not conform the surface and form islands with diameters of about 20 nm and separations of about 10 nm (Fig. 14.5). When placed over an Ag layer coated with a dielectric such as HfO2, the surfaces become super-absorbers, spanning a large wavelength range which covers the Raman excitation and scattering wavelengths (Fig. 14.5b). When a SEM picture (inset of Fig. 14.5c) is used to derive a model surface, electromagnetic full field calculations can be used to estimate hot spot density and enhancement factor for a given wavelength (Fig. 14.5d). It is seen that our surfaces generate SERS enhancements with of 10^9 around 700 nm (Fig. 14.5e) if the correct dielectric thickness is chosen. Hot-spot spacing of about 20 nm is achieved in such a configuration.

Note that, the image consists of signal from high enhancement regions (so called hot-spots) and improving hot-spot density results in a better sampling of the biomolecular structures. With sparse hot-spots, a continuous nanoscale map of the sample cannot be achieved. During the project we will investigate the effect of nanoscale non-uniformity on the imaging resolution while also looking for ways to reliably improve characteristic dimensions of the plasmonic structures towards 10 nm range.

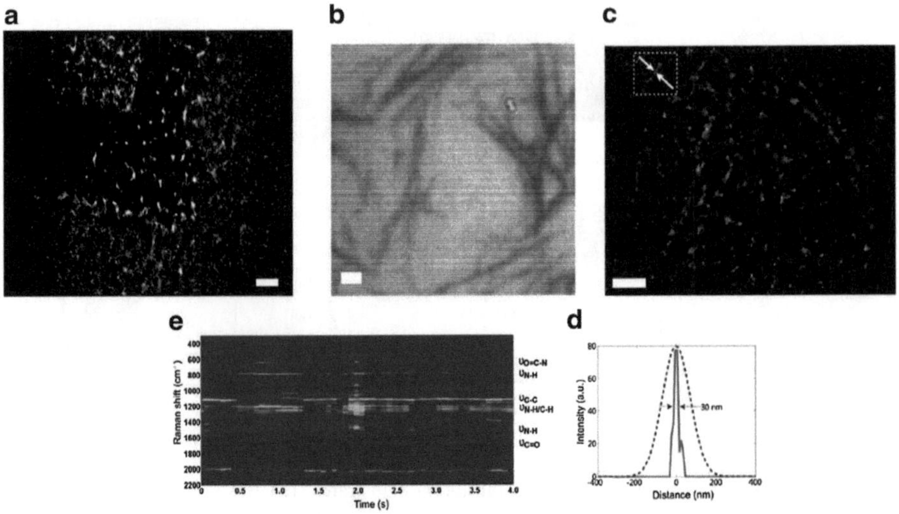

Fig. 14.4 Stochastic reconstruction SERS imaging of self-assembled peptide nanostructures: (**a**) Reconstruction of a peptide layer imprinted onto the plasmonic substrate using an elastomeric mold. (**b**) Bright field optical micrograph of a peptidic nanofiber network, directly self-assembled on the plasmonic substrate. (**c**) A wide-field SERS video sequence is recorded at the same location and resulting images are used to reconstruct image of self-assembled nanofiber network. Due to high SERS sensitivity, residual peptide presence on the surface complicates the resolution of individual peptide nanofibers. A well-separated, thin peptide nanobundle (highlighted by *arrows*) is used to plot the line profile. (**d**) The peptide bundle is observed as approximately 30 nm in diameter (*solid line*), a width smaller compared to the diffraction limited resolution (represented by *dashed line*). (**e**) During wide-field SERS video imaging of peptide nanofibers, SERS spectra was simultaneously recorded from a spot located at the center. Amide VI (630–750 cm^{-1}), Amide V (700–750 cm^{-1}), Amide III (1,230–1,300 cm^{-1}), Amide II ($\sim$1,550 cm^{-1}) and Amide I (1,600–1,700 cm^{-1}) bands associated with peptide backbone structure and side chain composition can be identified in the spectra. (δ, in plane bending vibration; ν, stretching vibrations.) (All scale bars are 1 μm wide). Again, whole integrated Raman spectra are used with a low sensitivity CMOS camera for the reconstructions. Due to lack of spectrally resolved optics during the collection of this data set, the rich spectral information can't be fully used to reconstruct a chemically specific HYPERSTORRM image

14.4 Feasibility of Stochastic Imaging Using the SERS Signal

We have made initial demonstrations of HYPERSTORRM imaging using a low sensitivity CMOS camera, which uses the whole integrated Raman spectrum, on two different types of samples as shown in Figs. 14.4 and 14.6: We have successfully imaged selfassembled peptidic nanofibers, as well as fibrillar structures of cardiomyocytes (rat heart cells) using highly-uniform SERS substrates with 20–30 nm resolution (Fig. 14.4d). The results are promising, however, in its present form, a nanometer resolution Raman map can't be generated as this requires EMCCD cameras and a multi/hyperspectral imaging system. EMCCD cameras provide about 100–1,000 times greater sensitivity compared to the CMOS sensors

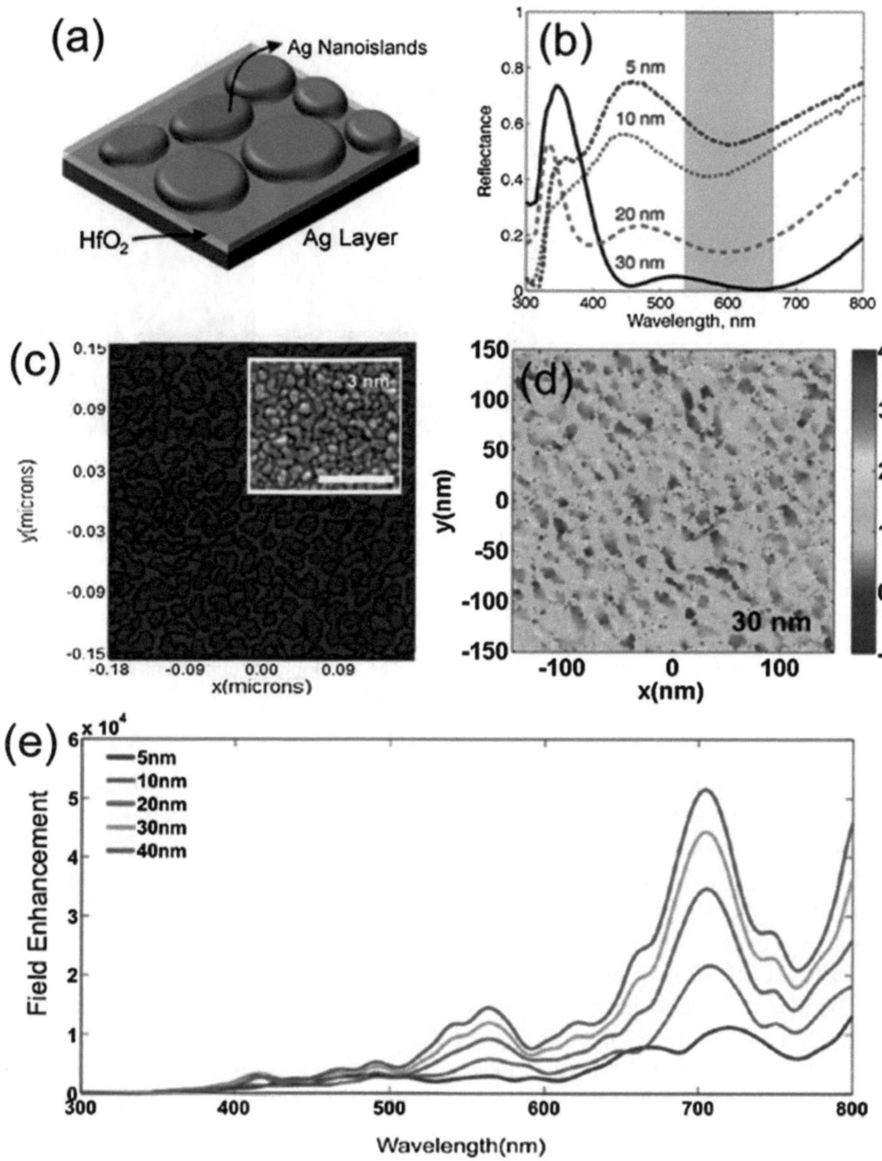

Fig. 14.5 Improved SERS substrates with single molecule sensitivity and high hot-spot density: (**a**) Ultrahin Ag films on dielectric layer on Ag substrate, segregates into nanoislands to form MIM surfaces. (**b**) These surfaces are omnidirectional superabsorbers if the dielectric thickness is around 30–40 nm. (**c**) A model surface is constructed for simulation using SEM micrographs (inset, scale bar 250 nm). (**d**) Full field simulations show $|E|^2$ field enhancement of 10^4 ($\log|E|^2$ is plotted). (**e**) Wavelength dependent field enhancement is shown for dielectric thicknesses of 5–40 nm. About 3×10^9 SERS enhancement is seen around 700 nm wavelength

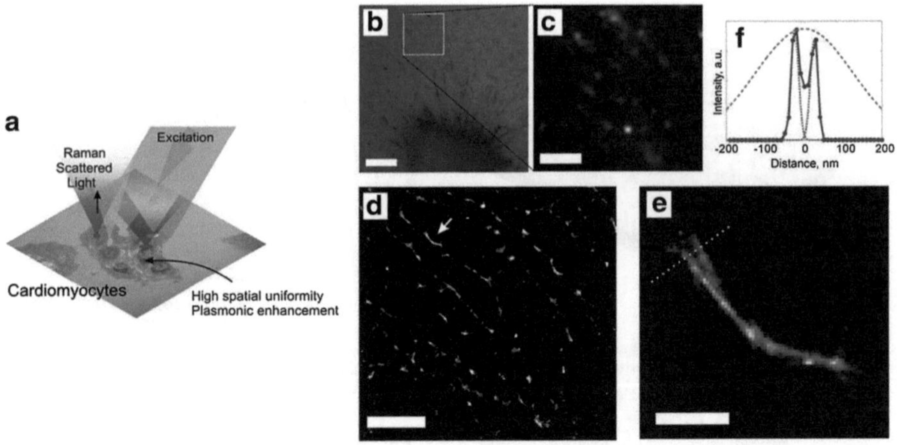

Fig. 14.6 Feasibility of SERS-STORM imaging. (**a**) Schematic description and (**b**) bright-field image of the edge of a cardiomyocyte on the plasmonic substrate. (**c**) A frame extracted from a video sequence of SERS images recorded using wide-field illumination and a low-cost CMOS camera, shows blinking SERS signal. (**d**) Stochastic reconstruction reveals fibrillar structure. (**e**) Detail of region shown by *arrow* in (**d**). (**f**) Line profile shows individual fibers with apparent diameter of 20 nm, separated by 50 nm. Diffraction limited profile is given by the *dashed line*. The reconstructed images use the whole integrated Raman spectrum, which can't currently be spectrally separated for chemically resolved HYPERSTORRM. The development effort aims to achieve such high resolution imaging, however with chemical specificity

used in acquisition of Figs. 14.4 and 14.6. EMCCDs have sufficient sensitivity which allow for the registration of signals from individual bands (which would contain 5–10 % of total optical power in a SERS blink). We demonstrate that the SERS signal can be acquired without optical damage to the chemical structure of biomolecular constructs (Fig. 14.6e). Therefore, although chemical resolution is not achieved in the preliminary results, they show that chemically specific HYPERSTORRM imaging is feasible.

The idea of hyperspectral imaging is tested as shown in Fig. 14.2 by using a one dimensional grating (as shown in Fig. 14.7). Clearly, one can separate the 0th order image from the 1st order image. In addition, when used with a SERS substrate as described earlier, blink events can be captured simultaneously in the 0th and 1st order images. Spectral separation of the content of the blink events are visible. However, for final implementation, a two dimensional optimized grating must be used with a high sensitivity EMCCD camera. Also mathematical methods must be fully developed so as to achieve highest signal to noise ratio and spectral clarity.

Within a first order semi-classical approximation, the SERS photon generation areal density rate $\varphi(x, y, \omega_s, t)$ observed on a plasmonic substrate is a function of a large number of variables, such as spatial orientation of fields, molecular configurations, and hot spots. Also, the SERS signal depends on Raman excitation and scattering frequencies, ω_{exc} and ω_s, as well as time t. The time dependence is observed to be strongly dependent on other parameters, such as excitation intensity,

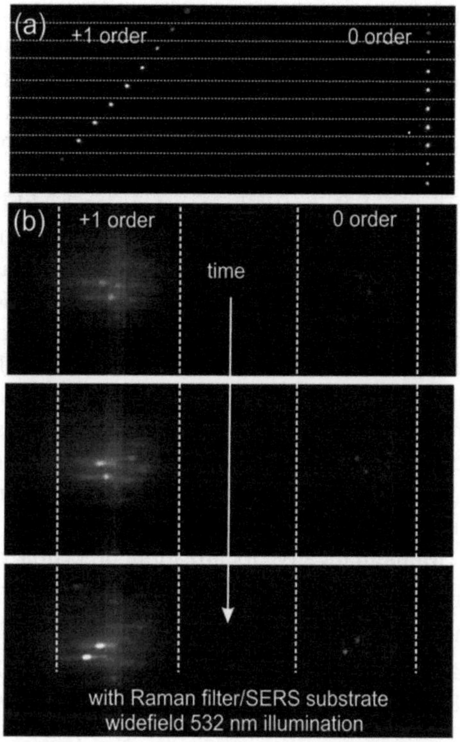

Fig. 14.7 Preliminary results on implementation of hyperspectral imaging of SERS signals. (**a**) Using a one dimensional transmission grating, 0th and 1st orders of a point source is imaged using the setup shown in Fig. 14.2a. The 0th order image does not shift with wavelength, whereas the 1st order image shifts due to diffraction. An RGB CMOS camera is used. (**b**) Consecutive frames are given from a video capture of SERS substrate treated with *methylene blue*, using a monochrome camera showing the 0th and 1st order regions. Again, the 0th order region exhibits discrete *blinking spots*, which can be used for localization. The 1st order region shows correlated spots, smeared in the *horizontal* direction due to spectral content of the SERS blinks. Initial results suggest that, using a high sensitivity EMCCD camera and an optimized 2D grating, more orders can be captured and be used for improved spectral/spatial resolution (Unpublished data)

local field enhancement factor, and chemical composition of imaged structures. At a given time t, the infinitesimal intensity $dI = \varphi(x, y, \omega_s, t)dA$ generated at an area dA at a location (x,y) from a molecular component i is given as [5].

$$dI_i\left(x, y, \omega_s, t\right) = QI_L\left(x, y, \omega_{exc}\right)\varepsilon(\omega_{exc})^{-1}\Omega \mathrm{S_i}\left(\omega_s, t\right) N_i\left(x, y, t\right) E_F\left(x, y, \omega_s, \omega_{exc}\right) dA \tag{14.1}$$

Where $\mathrm{S_i}\left(\omega_s, t\right) = \frac{d\sigma_i(\omega_s,t)}{d\Omega}$ is the differential scattering cross-section of molecular component i, characterizing its Raman spectrum. Here, Q is the overall optical collection and the detection efficiency; $I_L(x, y, \omega_{exc})$ is the excitation power density;

$\varepsilon(\omega_{exc})$ is the photon energy at excitation frequency; Ω is the solid-angle collected by the objective; $N_i(x, y)$ is the areal density distribution of molecular component I; and $E_F(x, y, \omega_s, \omega_{exc})$ is the function describing the spatial distribution of SERS enhancement factor. Assuming spatially uniform illumination, $I_L(x, y, \omega_{exc}) = I_{L0}$, a wide-field image can be formed using the SERS signal from a spatial pattern contained in the molecular layer $N_i(x, y)$ provided that the SERS enhancement distribution $E_F(x, y, \omega_s, \omega_{exc})$ is spatially constant or highly uniform. Although interference effects and speckle patterns are typically present during wide-field coherent illumination, uniform illumination assumption is reasonably satisfied for small dimensions. If the plasmonic modes responsible for Raman enhancement can be regarded as uncoupled or weakly coupled, hot spots can be thought of as operating independently. In such a case, regardless of the details of fluctuation mechanisms, the fluctuations originate from local effects which are uncorrelated. Such an optical signal, i.e. one with uncorrelated nanoscale spatial and temporal structure at the source, can be used for super-resolved stochastic imaging.

The simplest and most straight forward way of achieving multispectral STORM images is to reconstruct individual spectral channels separately and combine them by post processing. Mathematically, localization precision of point-like objects imaged in two-dimension is given by $\sigma_x^2 = \frac{r_0^2 + q^2/12}{N} + \frac{8\pi r_0^4 b^2}{q^2 N^2}$, where r_0 is the standard deviation of the point spread function, N is the total number of photons collected, q is the size of an image pixel, and b is the background noise per pixel [20]. For single blink events resulting in $\sim 10^5$ SERS photons, the resolution is expected to be on the order of few nanometers. It is not clear if the filtering required for spectral separation will allow collection of 10^5 SERS photons during a single blink event. Reduction of the collected photon number to 10^3 SERS photons will limit the resolution to tens of nanometers. It may be required to use modified algorithms during extraction of spot locations based on data from multiple cameras. Techniques alternative to finding the distinct individual centroid locations exist for super-resolution imaging such as optical fluctuation imaging (SOFI) [21]. Inspired by such techniques, we propose to use a correlation based technique, not for centroid localization, for hyperspectral image separation.

One of the novelties in this work is the use of a two dimensional grating for hyperspectral imaging as shown in Fig. 14.2c. A two dimensional grating positioned in front of a camera will cause the image to split into 0th and higher order mode images. The higher order mode images will be spectrally separated. In the imaging configuration shown in Fig. 14.2, the 2D dispersion of the grating is characterized for orders m and n by a shift of the intensity of the spectral components of the 0th order image by $\alpha m\lambda$ and $\alpha n\lambda$ in x and y directions. The collection quantum efficiency and diffraction efficiency are assumed to be uniform. Also, the presence of the 2D grating is assumed not to cause distortions in higher order mode images, which is reasonable assumption for small diffraction angles. For the imaging configuration shown in Fig. 14.2d, the intensity distribution observed by the camera can be approximated as

$$I(x,y,t)=\sum_{m,n}\gamma_{m,n}\iiint_{x0,y0,\lambda}\delta(x-x0-\alpha m\lambda)\,\delta(y-y0-\alpha n\lambda)\,S(x0,y0,\lambda,t)\,d\lambda dx0dy0 \tag{14.2}$$

Where $\gamma_{m,n}$ is the efficiency of the order, and m,n, $S(x0,y0,\lambda,t)$ is the source located at (x0,y0) with spectral distribution given by a function of λ and a time dependent intensity fluctuation. The diffraction limit can be incorporated by including a point spread function (PSF) to replace the delta functions $\delta(x)\delta(y)\overset{\Delta}{\rightarrow}PSF(x,y)\triangleq exp\left(-\frac{x^2+y^2}{2\sigma^2}\right)$. The intensity of the spectral function $S(x,y,\lambda,t)$ can be written as the product of a time independent spectral function $S(x,y,\lambda,)$ that depends on location (x,y), wavelength λ, and a time and location dependent random variable $T(x,y,t)$, as $S(x,y,\lambda,t)=S(x,y,\lambda)\,T(x,y,t)$. This can be justified by the experimental results shown in Fig. 14.7, where simultaneous measurement of spectral fluctuations and location dependent blinking events are recorded on a SERS substrate. In other words, we ignore small fluctuations of spectral positions of the Raman bands, and treat the blinking of the SERS signal as a temporal envelope that modulates the full SERS spectrum. Because the camera reads intensities that are absolute positive, it is more convenient to work with the time derivative of the intensity $\dot{S}(x,y,\lambda,t)=S(x,y,\lambda)\,\dot{T}(x,y,t)$. We assume that $\langle\dot{T}(x,y,t)\rangle=0$, which says that the time average of the fluctuations is zero and $\langle\dot{T}(x,y,t)\,\dot{T}(x',y',t)\rangle=\delta(x-x',y-y')$, which explains that different locations are uncorrelated in time. We don't make any assumptions about the autocorrelation function of $\dot{T}(x,y,t)$, apart from the observation that the autocorrelation has a finite bandwidth and blink events occur with an average frequency on the order of a second.

We can write the intensity separated in orders (m,n) as

$$I^{m,n}(x,y,t)=\gamma_{m,n}\iiint_{x0,y0,\lambda}\delta(x-x0-\alpha m\lambda)\,\delta(y-y0-\alpha n\lambda)\,S(x0,y0,\lambda,t)\,d\lambda dx0dy0 \tag{14.3}$$

Then the 0th order intensity is integrated over the spectrum, $I^{0,0}(x,y,t)=\gamma_{0,0}\int_\lambda S(x,y,\lambda,t)\,d\lambda\triangleq\gamma_{0,0}S(x,y,t)$. The intensity for (m, n) = (1,0) will be $I^{1,0}(x,y,t)=\gamma_{1,0}\int_\lambda S(x-\alpha\lambda,y,\lambda,t)\,d\lambda$. The problem is to determine $S(x,y,\lambda)$. We assume that the fluctuations from different molecules are uncorrelated, i.e. use $\langle\dot{T}(x,y,t)\,\dot{T}(x',y',t)\rangle=\delta(x-x',y-y')$, and get

$$\int_t\dot{S}(x,y,\lambda,t)\,\dot{S}(x',y',\lambda,t)\,dt=\delta\left(x-x',y-y'\right)S(x,y,\lambda) \tag{14.4}$$

Where $S(x,y,\lambda)$ is the time average spectrum of molecule at position (x,y). Using a similar approach, we can calculate the correlation of pixel in the 0th and 1st order image regions, to find $\int_t \dot{I}^{0,0}(x,y,t)\,\dot{I}^{1,0}(x',y',t)\,dt = \delta(x-x'+\alpha\lambda)\,\delta(y-y')$ $S(x,y,\lambda)\int_\lambda S(x',y',\lambda)\,d\lambda$. Therefore, the correlations of pixels (x, y) in the 0th order image with the $(x',y') = (x-\alpha\lambda, y)$ in the (1,0) order image result in intensities proportional to $S(x,y,\lambda)$ the λ spectral component of the corresponding source pixel at (x,y). A similar approach can be used to improve convergence of the correlations using other orders. A more detailed formalism must be developed if the Raman band frequencies are assumed to fluctuate as well as the intensities.

14.5 Conclusions

In summary, we discuss a novel form of label-free super-resolution imaging, namely hyperspectral stochastic optical reconstruction Raman microscopy, which has the potential to provide chemically resolved Raman maps with nanometric resolution. The technique relies on SERS, and has inherent limitations: The resolution is limited by hot-spot spacing of the plasmonic surfaces, and observations are limited to the surface plane of the plasmonic substrates. Still, the method can be used to study biologically relevant specimens, such as membrane phenomena.

Acknowledgments The author acknowledges TUBITAK Grant 111M344 and EU-FP7 NanobacterphageSERS for financial support.

References

1. Wang MC, Min W, Freudiger CW, Ruvkun G, Xie XS (2011) RNAi screening for fat regulatory genes with SRS microscopy. Nat Methods 8:135–138
2. Steidtner J et al (2008) Tip-enhanced Raman spectroscopy and microscopy on single dye molecules with 15 nm resolution. Phys Rev Lett 100:236101
3. Jeanmaire DL, Van Duyne RP (1977) Surface Raman spectroelectrochemistry: part I. Heterocyclic, aromatic, and aliphatic amines adsorbed on the anodized silver electrode. J Electroanal Chem Interfacial Electrochem 84:1–20
4. Albrecht MG, Creighton JA (1977) Anomalously intense Raman spectra of pyridine at a silver electrode. J Am Chem Soc 99:5215–5217
5. Stiles PL, Dieringer JA, Shah NC, Van Duyne RR (2008) Surface-enhanced Raman spectroscopy. Annu Rev Anal Chem 1:601–626
6. Nie S, Emory SR (1997) Probing single molecules and single nanoparticles by surface-enhanced Raman scattering. Science 275:1102–1106
7. Kneipp K et al (1997) Single molecule detection using surface-enhanced Raman scattering (SERS). Phys Rev Lett 78:1667–1670

8. Kneipp K, Wang Y, Dasari RR, Feld MS (1995) Approach to single molecule detection using surface-enhanced resonance Raman scattering (SERRS): a study using Rhodamine 6G on colloidal silver. Appl Spectrosc 49:780–784
9. Kneipp K et al (1998) Surface-enhanced Raman scattering (SERS) – a new tool for single molecule detection and identification. Bioimaging 6:104–110
10. Le Ru EC, Etchegoin PG (2012) Single-molecule surface-enhanced Raman spectroscopy. Annu Rev Phys Chem 63:65–87
11. Rust MJ, Bates M, Zhuang X (2006) Sub-diffraction-limit imaging by stochastic optical reconstruction microscopy (STORM). Nat Methods 3:793–796
12. Stranahan SM, Willets KA (2010) Super-resolution optical imaging of single-molecule SERS hot spots. Nano Lett 10:3777–3784
13. Cang H et al (2011) Probing the electromagnetic field of a 15-nanometre hotspot by single molecule imaging. Nature 469:385–388
14. Balzarotti F, Stefani FD (2012) Plasmonics meets far-field optical nanoscopy. ACS Nano 6:4580–4584
15. Perney NMB et al (2006) Tuning localized plasmons in nanostructured substrates for surface-enhanced Raman scattering. Opt Express 14:847–857
16. Ayas S et al (2012) Raman enhancement on a broadband meta-surface. ACS Nano 6:6852–6861
17. Ayas S, Cinar G, Dana A et al (2013) Label-free nanometer-resolution imaging of biological architectures through surface enhanced Raman scattering. Sci Rep 3:2624
18. Ayas S et al (2014) Counting molecules with a mobile phone camera using plasmonic enhancement. ACS Photonics 1:11
19. Fisher J, Baumback MM, Bowles JH, Grossmann JM, Antoniades JA (1998) Comparison of low-cost hyperspectral sensors. In: Proceedings of SPIE 3438, Imaging Spectrometry IV, 23. doi:10.1117/12.328112
20. Hess ST, Girirajan TPK, Mason MD (2006) Ultra-high resolution imaging by fluorescence photoactivation localization microscopy. Biophys J 91:4258
21. Dertingera T, Colyera R, Iyera G, Weiss S, Enderlein J (2009) Fast, background-free, 3D super-resolution optical fluctuation imaging (SOFI) PNAS 106 52 22287-22292

MIX
Papier aus verantwortungsvollen Quellen
Paper from responsible sources
FSC® C105338

If you have any concerns about our products,
you can contact us on
ProductSafety@springernature.com

In case Publisher is established outside the EU,
the EU authorized representative is:
Springer Nature Customer Service Center GmbH
Europaplatz 3, 69115 Heidelberg, Germany

Printed by Libri Plureos GmbH
in Hamburg, Germany